The Coal Mines of Buxton

Alan Roberts & John Leach

Scarthin Books
Cromford, Derbyshire
2008

The cover illustration is taken from an aquatint of 1796,
A View of Buxton, by William Martin.
Reproduced by kind permission of Derbyshire Museum Service.

Published 1985
by Scarthin Books, Cromford, Derbyshire DE4 3QF
Reprinted 2008

www.scarthinbooks.com

e-mail: clare@scarthinbooks.com

Typesetting and layout by CJWT Solutions, St Helens WA9 4TX

ISBN 978-0-90775-810-5

Printed by Lightning Source

CONTENTS

Frontispiece: Pit drainage sough in the Goyt Valley.

CHAPTER ONE

COAL MINES IN BUXTON?

About two hundred years ago the fifth Duke of Devonshire was developing Buxton as a fashionable spa town. The Crescent was opened in 1784, and by 1790 the Buxton season had become well established.

A curious visitor to Buxton in 1790, seeking a short excursion by riding out towards Macclesfield, would have encountered unusual activity near the Goslin Bar tollgate. Down Level Lane a small line of carts was being filled with coal from a large stack. From the mouth of a nearby tunnel a boat laden with coal appeared from time to time, to be unloaded, and returned into the darkness, propelled by a boy of about twelve.

Continuing along the turnpike to the ridge, our visitor would have seen, off to his left, the horse engines at the Old Rise Pitt, and the New Rise Pitt. The horse was led by another young lad, round a circular track, first one way and then the other. With a steady, purposeful rhythm every two or three minutes another full corf of coal appeared at the pit top, and the waiting carts were filled. Down by the Goyt River there was a similar scene, but instead of a horse engine the coal was being wound up one of the many shafts by two men, toiling at a simple two-handled windlass.

Unless he had set out early in the morning or in the evening he would not have seen any of the colliers themselves, for they were busy underground. Returning to Buxton, he would only be reminded of all this activity by meeting empty coal carts returning to the workings. He would pass a few isolated cottages, and then the open fields, before reaching Buxton itself. Once in the town, apart from the occasional cart making coal deliveries to the inns and hotels, all signs of the collieries disappeared, and our visitor's thoughts would turn to the evening's round of pleasure.

Unless he had bought a season ticket, he would have to pay three shillings for admission to the Card Room.[1] To earn that much the boy, George Turner, would have had to drive the engine horse round and round on its circular track for six whole days. The price of a ticket to a Ball was seven shillings. For this, Thomas Goodwin and James Sutton together would have had to wind 560 corves of coal, each weighing one hundred weight, up a twenty-yard-deep shaft.

Of course, our curious visitor would never have thought of making any comparison between the price of his enjoyments and a collier's labour, and most of the gentry in Buxton for the Season were not even aware that any such labour existed within a few miles of their own glittering society.

Air shaft of the Dane Colliery, plainly visible from the Buxton to Congleton road.

Equally, today, Buxton is not normally thought of as a centre of the coal mining industry; indeed the town would probably prefer its reputation to be based on its history as a spa and on its location in a major open air recreational area. The residents of Burbage, the western suburb of Buxton, probably think of it as a pleasant residential area rather than as a former mining village (which, to a limited extent, it was).

Nevertheless, there were coal mines in the hills above Burbage. Although insignificant on a national scale (the total extraction of coal was probably less than the largest modern coal mine can produce in one year), the mines were important locally and, before the improvement of rail connections to the town, played a significant part in the development of Buxton as a major centre of the limestone extraction and limeburning industries.

The coal mines therefore represent an important element of local history and one with many interesting features as well. This short book sets out something of the development of the mines and puts them in a context of the development of local trade and communication. The account is inevitably incomplete; the last official mining of coal took place in 1919 - it may be that there are documents or photographs relating to the mines tucked away in local drawers or cupboards - we would be glad to see any such material.

Strictly speaking, the mines were not in Buxton. They were in the Parish of Hartington Upper Quarter, about 2 miles from Buxton, but the description of them as 'Buxton coal mines' provides an easier reference point for those unfamiliar with local parish boundaries. The mines also had different names at different times, as the workings developed away from their point of origin.

At different times (or on different documents) they are referred to as: Goyt Colliery, Goyte's Moss Colliery, the Level Mine, Buxton Collieries, Burbage Colliery, Thatch Marsh Colliery, Axe Edge and Thatch Marsh Colliery.

Essentially, all these names refer to all or part of the same set of workings, which are referred to here as Buxton's Coal Mines.

The coal mines close to Buxton were fairly modest affairs, with mining activity from about 1600-1919. The most productive period was from about 1780-1880 and in the 1790-1810 period they were fairly substantial operations for that time. Due to the lack of extensive reserves, they failed to develop during the Victorian era (to the great benefit of the local scenery) but continued to operate until the exhaustion of the readily available reserves in 1919.

The development of these mines is inextricably linked with the growth of the local lime-burning industry.

Local geology, the demand for lime, and the communications available for importing better quality coal and exporting lime were all important factors in affecting the production of local coal which are considered in Chapters 2-4.

The physical details of the coal mining operation, in terms of the nature and extent of the workings and their estimated production, are described in Chapters 5 to 7, while the available information on the miners is presented in Chapter 8.

The final two chapters deal with some of the more obscure local connections with coal mining and with a description of the visible remains of the mining activities. Chapter 9 also includes a small section on the small scale lead and barytes mines in the Grin area which are not recorded elsewhere.

Although maps and diagrams are given throughout the book, the Ordnance Survey 1:25000 map of the White Peak is a useful additional aid and all grid references used can be found on this map.

Footnote references for the whole book follow chapter ten. Appendix 6 at the rear of this book contains useful conversion tables.

CHAPTER TWO

THE COALFIELD

The history of coal mining in Britain can be traced back to at least 1100 A.D. by documentary references, but the emergence of coal mining as a major industry becomes apparent in Tudor times.

Prior to that, wood was the major source of fuel and industries such as the smelting of iron and lead, with known origins back to Roman times and beyond, relied on wood charcoal as a fuel for smelting. Nef[1] traces the factors that caused the more rapid growth in coal production in the 16th and subsequent centuries. A growing population with a higher demand for food caused pressures on woods and forests, which were progressively cleared for agricultural purposes; demand for wood for ship building, house building and charcoal production exceeded the available supply; importation of timber became necessary. These factors created the right conditions for a rise in supply of an alternative fuel. Nef estimates that between 1451 and 1641 the price of firewood increased by a factor of 8, where general prices increased by a factor of only 3. There were large local variations; the Earl of Rutland's accounts[2] record that a load of charcoal cost 3/4 in 1536 and 25/- in 1586. Nef estimates British annual coal production (tons) to have been as follows:

	1551-1600	1681-1690	1781-1790	1901-1910
Midlands (including Yorkshire, Lancashire, Cheshire, Derbyshire, Shropshire, Staffordshire, Nottinghamshire, Warwickshire, Leicestershire, Worcestershire)	65,000	850,000	4,000,000	100,180,000
Britain	210,000	2,982,000	10,295,000	241,910,000

Thus, during the 16th and 17th centuries, coal was becoming the main fuel in coal mining districts but its costs to the consumer were heavily dependent on transport costs. Nef[3] comments that generally the cost of land transport prevented any considerable sale at distances of more than ten or, at the most, fifteen miles from the collieries. Coal transported ten miles by land would cost four times as much as at the pithead. Transport by sea or river was much less costly and in 1675 the cost of carrying coal three hundred miles by water was only the same as the cost of transport for 15-20 miles by land.

There are early records of the use of coal in limeburning and the significance of its availability. Limeburners in Southwark were, in 1307, forbidden to use coal because of the fumes produced. In the Middle Ages, coal was used when available,for limeburning for major building works such as Cathedrals. During the reign of Henry VIII, a 'want of sea coles' is said to have caused a decline in the standard of husbandry (i.e. agricultural production) in Cambridgeshire because of a resultant shortage of lime. As early as 1640 a shortage of coal caused a serious crisis in the building trades of London, again through a shortage of lime. Thus the two traditional uses of lime, in agricultural and in construction work, became dependent on the availability of coal supplies and it is estimated that by the end of the 17th century limeburners used tens of thousands of tons of coal annually.

The bounds of what is often known as the 'Cheshire' coalfield are roughly Mossley in the north and the Roaches in the south, Buxton to the east and Macclesfield to the west. Coal has been extracted from the whole length of the coalfield which includes parts of the former counties of Cheshire and Lancashire to the north and the present counties of Cheshire, Derbyshire and Staffordshire in the central and southern areas.

Within Derbyshire, coal has been mined at Chisworth and Ludworth, New Mills and Rowarth, Ollersett and Chinley, Buxworth and Beard and at Whaley Bridge and Axe Edge. Towards the end of the last century 'The Shallcross Colliery Company' was a large concern in the Whaley Bridge area.

In different parts of the coalfield various types of coal were mined depending upon the proximity to different geological horizons. These different coals had a host of local names and within a short distance the same seam often changed its name. For example the 'Yard' coal was called 'Goyt' on Axe Edge, 'Kiln' at Whaley Bridge and 'Mountain' at Mellor. For the purposes of this chapter, only the current geological names are used.

The 'Ringinglow' coal is the lowest horizon of coal and so John Farey[4] in his survey of 1811 called it the 'first coal'. It was mined extensively on the Hallam moors above Sheffield, from where it gets its name, but, with the exception of the Axe Edge mines, very little was mined on the western side of the Peak.

Slightly above the 'Ringinglow' horizon lies the Simmondley coal. It was not mined to any great extent and, incidentally, outcrops in Crackenedge Quarry above Chinley.

The next horizon and that most extensively mined is the 'Yard' coal, which Farey called the 'second coal'; thick and dirty, it was used mainly for the burning of lime and so assumed the name 'Kiln Coal' in the Whaley Bridge area where a seam of nearly two metres thick existed in the Shallcross colliery.

'Ganister' coal is almost absent except for small undertakings around Chisworth. However, its seat earth was at one time extracted for refractory manufacture at Fox Hill and south-west of Saltersford Hall in the Toddbrook area.

The next horizon 'White Ash' coal, was mined substantially at Whaley Bridge and Shallcross. It was also mined at Furness Vale where its seat earth was mined until 1964 for the manufacture of firebricks and backs.

'Red Ash' coal workings are numerous but of small extent, probably due to its sulphurous nature. It was often mixed with 'Yard' coal for limeburning.

The highest coal in the district is the 'Big Smut'. Due to its variable nature this has not been mined extensively except around Ludworth Moor. A single seam with its associated fireclay was mined until recently at Pott Shrigley.

Although extraction has virtually ceased in this coalfield, considerable areas of unworked coal still remain.

The description of the many collieries and small mines which existed is beyond the scope of this book, but a few of the more notable ones and those closest to Buxton can be mentioned.

Those of the Whaley Bridge area were perhaps the most extensive. Mention has already been made of these and it is from these we get one of our earliest references. In 1606[5] a dispute over the ownership of a coalmine at Fernilee was referred to the 'Star Chamber' court. Thomas Bagshawe of Ridge Hall alleged that Ralph Cooke, yeoman, and twenty-two other persons, armed, threatened his servants if they continued to mine. Bagshawe also accused them of assaulting John Davie 'a collier expert'.

A phenomenon of the 'Blackclough' (or 'Beat') Derbyshire mine, was that the workings were mined directly through a geological fault into a different seam. The workings of this mine connected with those of the 'Dane', Cheshire, which produced some of the cleanest and best quality coal in the area out of the 'Ringinglow' seam which reached four feet thick.

Mention should also be made of a small outcop of 'Simmondley' coal which was mined at Errwood. The mines here served the 'Hall', the local community and possibly the gunpowder and barytes works. One of the mines, reputed to be a mile and a half long, is now flooded by Errwood reservoir. Another, in Shooters Clough, utilised a narrow gauge tramway.

Other small workings close to Buxton which merit recording are on Ladder Hill, in Wildmoorstone Clough and below Ladbitch wood.

One local working was noted by Farey as: 'Combes Moss nearly north of Buxton, a table mountain of Shale and first Grit, has a small patch or depressed hummock of the first coal, not denudated or stript off ....',

This is unusual because in his survey[6] he lists over 500 collieries in the county and adjacent areas, and, within that list records Combes Moss as 'Formerly', i.e. not worked. This could mean that the mine closed between the original and second publications (1811 and 1815). Other mines listed, 'Formerly', are Chest, Ferneylee, Gap Sitch, Hay Clough, Latch, Mousetrap, Notbury, Quarnford and Robinsclough.

Appendix 1 lists extracts from 'Farey' relating to the southern part of the coalfield. The full list also records the type of coal extracted.

The object of the book is, however, to give as full an account as possible of the collieries and mines immediately to the west of Buxton, in the southern part of the coalfield. Before such an account can be given, a brief word must be said about the local geology of the area under review.

Buxton lies on the western arm of the great anticline that forms the 'Derbyshire Dome'. Eastwards from its axis at Woodale the full succession of Carboniferous limestones and succeeding rocks are evident. Westward, however, there is an abrupt change at Buxton, with the Namurian (sometimes called 'Millstone Grit') series of rock and shale lying unconformably on the 'Millers Dale' beds of Carboniferous limestone. Further west, there are a number of subsidiary folds in the Namurian series and it is within one of these folds, the 'Goyt Syncline' that the mines involved are situated. It must be noted that the angle of dip of the eastern arm of the syncline is greater than that of the western arm and by virtue of the fact that the coal seams lie within the syncline they will outcrop on opposite sides of same.

The coal, however, is of poor quality and the mines difficult to work, with faulting and flooding.

Short[7], in 1734, described the mines in the following terms:

'below shales, beds of ironstone, then a seam of coal about five feet thick',

'with much sulphur and brazil-iron pyrites',

'dipping about one yard in five'

'The upper coal was soft and flaky, fit only for burning limestone, the lower was harder but of indifferent quality'.

Outcrop in the 'Goyt' seam near Derbyshire Bridge.

'At about 150 feet depth there was a thick bed of blue clay, part of it ochreish, containing black lumps like rusty iron, with green copper in it'.

'Occasionally a vein of lead ore crossed the seam of coal'.

'The water in the pits was so extremely cold that miners who worked long there were in danger of losing their limbs'.

Watson[8] in a cross section of the Peak District, records the presence of a colliery on Combs Moss, and of the coal seams he numbers 'XX' and 'XXII' he says:-

XX Rock Coal: abounding with sulphuret of iron in its laminae, a bed about twenty two inches thick'.

'XXII Abounding with nodules of sulphuret of iron, termed Brasses, Bats etc., used in making sulphate of iron and carbonate of iron in thin laminae, crystallised in rhombs. In this stratum at Thatch Marsh near Buxton, veins of Sulphure of Lead are found in faults having Coal attached on both sides accompanied with sulphuret of iron'.

The coal seams involved are the 'Ringinglow' and the 'Yard' respectively.

The 'Ringinglow' coal which lies above the 'Chatsworth Grit' is the better quality of the two, and reached a thickness of 4 feet in the Axe Edge workings.

The 'Yard' coal lies above the 'Woodhead Hill Rock' and is usually shaly and sulphurous. The seat earth between the 'Woodhead Hill Rock' and the 'Yard' coal varies in depth up to four feet thick and is sandy, especially in the lower part. It has not been worked to any extent.

Details of the local coal seams are shown in Figs 1 and 3. The outcrop of the 'Ringinglow' seam runs southward from an east/west fault on the line of Berry Clough; it continues in this direction beyond the area covered by the diagram.It is intersected by two other east/west faults but is not greatly displaced at these. The 'Ringinglow' seam is missing for about one mile northwards from the Berry Clough fault and outcrops again near Wildmoorstone Brook. It dips from the outcrop at about 1 in 6 in a westerly direction.

The outcrop of the 'Yard' seam is more complex; it is basically an oval 1¼ miles x 1 mile with displacements due to several faults. This seam outcrops on either side of the Goyt Valley forming a shallow bowl from north of Derbyshire Bridge to Tinkerspit Gutter. A detailed cross section of the strata from the Cat and Fiddle to Ladmanlow is shown in the Geological Survey of Great Britain, Sheet SK07.

The above names are the modern names used by the Geological Survey. The mining records use other names. The Ringinglow Seam was referred to as the Mountain Seam about 1880-1900, and as the House Coal Seam

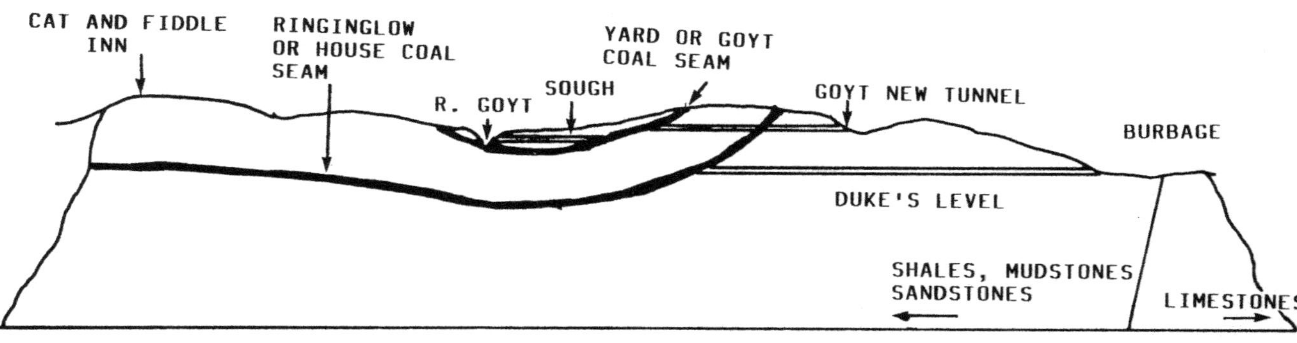

Fig. 1. Cross section of strata from 'Cat and Fiddle' to Burbage showing approximate position of coal seams and major underground features.

KEY

Tunnel	**Approximate Date of Construction**	**Height Above Sea Level (ft.)**	**Purpose**
Duke's Level	1770	1170	Access, canal transport, drainage
Goyt New Tunnel	1812	1360	Access, rail transport
Sough	?	1285	Drainage

No coal was mined below the level of the Duke's Level

prior to that. The Yard Seam was known locally as the Goyt Seam. Throughout the remainder of this book the traditional names of the House Coal Seam and Goyt Seam will be used. Both of these seams were recorded as 4 feet thick with a narrow dirt band at mid height.

The book will concentrate on the Goyt and Thatch Marsh collieries, in the area marked on Fig. 3.

Thus the area considered is the stretch of uninhabited moorland lying between the Cat and Fiddle Public House in the west, Berry Clough in the north, Axe Edge in the east and the Derbyshire/Staffordshire border in the south. Occupied by a few isolated cottages at one time, the area is now the province of sheep, grouse and walkers, apart from the traffic along the main roads that intersect it. Despite the traffic, the area still retains a wild and remote look. Lying between 1300 feet above sea level in the Goyt Valley and 1800 feet above sea level on Axe Edge, it is exposed to high winds, high rain and snowfalls and extensive snow drifting. Even today, with all the horsepower available on modern snowploughs, the road past the Cat and Fiddle has a certain notoriety under snowy conditions and is relatively easily blocked. One can feel isolated from the conditions inside a modern well heated car but to walk across these moors in adverse weather conditions is a salutary reminder that, in addition to the normal rigours of coal mining, men worked in these difficult conditions in one of the most exposed industrial working locations in England.

CHAPTER THREE

LOCAL COMMUNICATIONS

Like any other commercial product, coal needs to be produced and transported to its point of end use at a competitive price. Transport costs are a very important part of the final cost of coal, particularly in areas where communications are fairly poor, like the Peak District of two hundred years ago.

The communications system serving Buxton's coal mines therefore played an important part in determining the size of the market for the coal produced, hence for the investment back into the mines. Various aspects of this system are shown in Figs 2 and 3.

The earliest form of transport for coal, a relatively bulky product, was by packhorse over rough tracks worn by the passage of the horses themselves. A fully laden packhorse could carry up to about 2½ cwt of load (8 loads to the ton) and there are several accounts of the strings of packhorses, twenty or thirty or more, to be seen crossing the moors around Buxton. On the relatively thin soils lying on a hard rock base, as in the limestone country, the packhorse trails left relatively faint traces. However, on the steep slopes of the sandstones and shales, often of quite low strength and often covered by a thick layer of peat, the packhorses wore deep furrows into the ground which would tend to be enlarged by water draining down them. These hollowed out tracks, or holloways, remain today in many places and packhorse trails to the Buxton coal mines can easily be traced.

A packhorse route[1] from Buxton west towards Macclesfield is recorded as far back as 1600 in a manuscript of Anthony Bradshaw of Duffield, describing 'The Bounds, Extremities and Meres of the High Peake Forest'. Part of the southern boundary was defined as 'ffollowing Wye to Wyehead ffolowing Jaggers Gate to Goyte Water'.

Prior to 1759, one road and four main holloways served the coal field in the area under consideration (shown in Figs. 2 and 3) and some supplementary evidence about their age and continuations is available in each case.

A very early route is that from Ladmanlow passing to the south of the 'Terret' and running west to SK0330.7135. Along the course of the present A537 to SK0310.7135 where it ran to the north of the same to

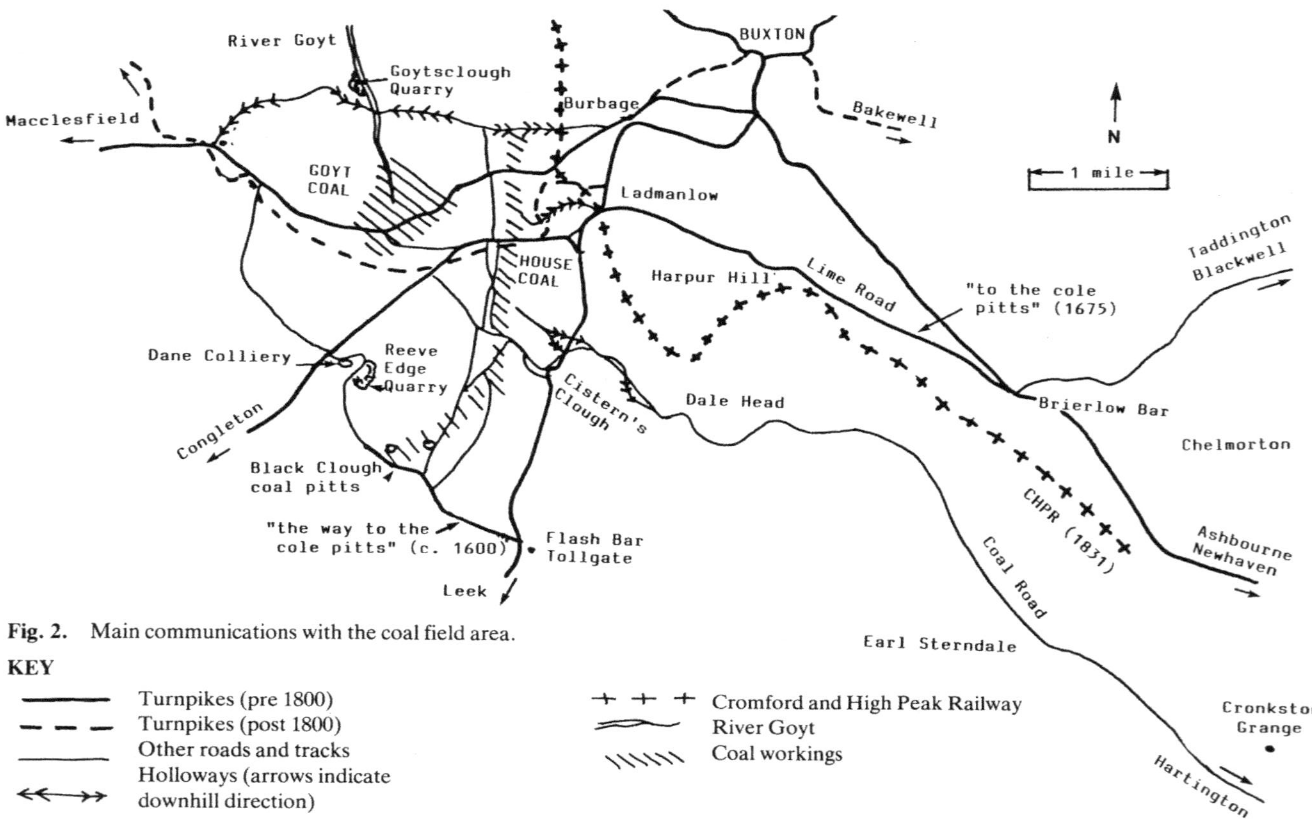

Fig. 2. Main communications with the coal field area.

KEY

Turnpikes (pre 1800)
Turnpikes (post 1800)
Other roads and tracks
Holloways (arrows indicate downhill direction)
Cromford and High Peak Railway
River Goyt
Coal workings

SK0262.7137, former junction with the Congleton road. From here it continued to join the present road at SK0192.7128. Most of the route can still be traced.

One holloway can be seen starting at the boundary of the limestone and the sandstone/shale at Ladmalow. It climbs the northern side of Ladman's Low (shown on modern maps as The Terret) as a holloway up to 18 feet wide at the top and 6 feet deep in places. The plantation on the top of Ladman's Low was planted after the holloway was formed; the holloway reappears on the far side of the plantation and descends the hill steeply to the valley of a small branch of the River Wye, crossing the A537, Buxton to Macclesfield road. It then climbs up the hill beyond the river and its route can be traced on the flatter ground where the coal seam outcrops. It seems likely that this is the route 'to the cole pitts' shown at Brierlow Bar on the 1675 map of Ogilby, described in more detail in Chapter 5. The route from Brierlow Bar to Ladmanlow at one time known as the 'Lime Road' was turnpiked in 1773 and the holloway described above appears to be a direct continuation of this route. Further evidence is contained in an Ordnance Survey map of 1879 on which a road (sometimes shown as Lydgate Lane) running from near Brierlow Bar towards Taddington past Farditch Farm is referred to as 'Old Coal Pit Lane'. One can therefore trace a continuous coal route from near Taddington to the coal outcrop.

Another holloway running from Burbage is no longer clear to the east of the Cromford and High Peak Railway although it probably ran along the course of the later 1759 Buxton-Macclesfield turnpike as far as the crossing of the infant River Wye. Beyond the stream it diverged north from this later turnpike, and west of the railway it is apparent as deep holloways climbing Burbage Edge about three hundred yards from the turnpike. These holloways pass through Burbage Plantation but are now blocked by a stone wall erected as a consequence of the Hartington Enclosure Award 1804. Beyond the wall the route continues as a public footpath along Berry Clough to the Goyt Valley. It then climbs the west side of the upper Goyt Valley in a holloway towards the plantation, where it rejoins the public footpath which leads behind Goytsclough quarry towards Stake Side, Stake Farm and a deep holloway section to the site of Stonyway Toll Gate (now occupied by the Shining Tor restaurant) where it rejoins the turnpike.

There is evidence for an earlier route along the line of the above 1759 turnpike between Burbage and the coal workings. Immediately to the west of Cromford and High Peak Railway a footpath leaves the turnpike and heads across to join the above route (instituted when the above route was blocked?). For about two hundred yards this footpath runs along a stone surfaced track and where the public footpath veers north this track

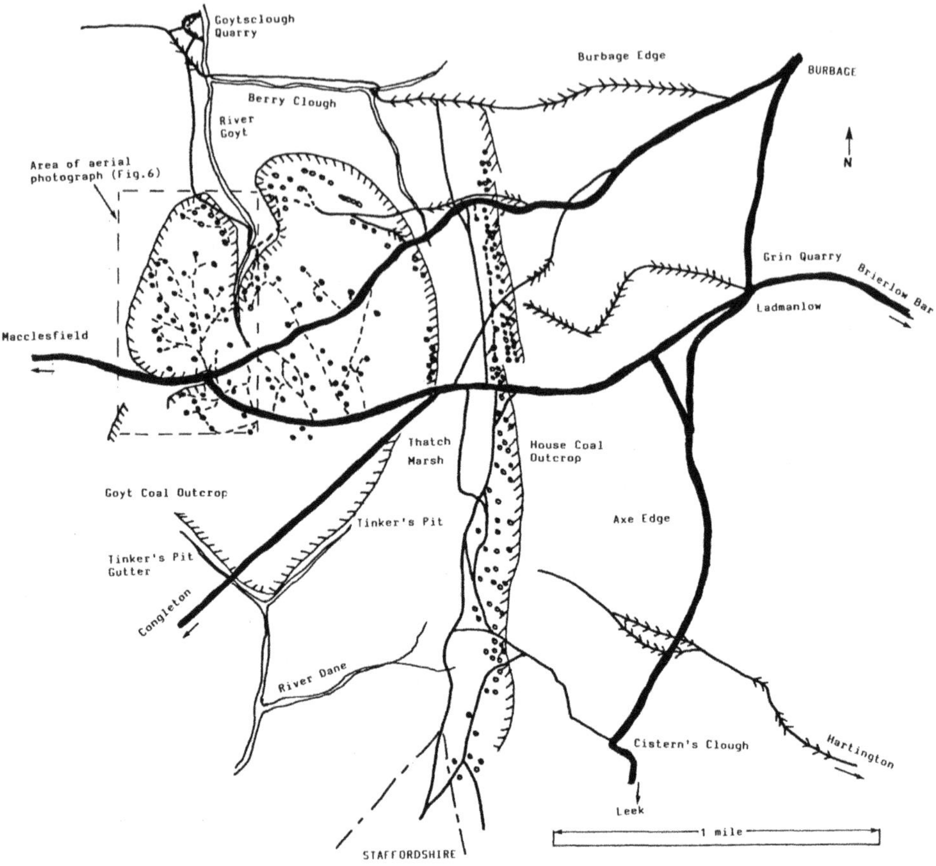

Fig. 3A. Early coal workings and road network c. 1790

KEY

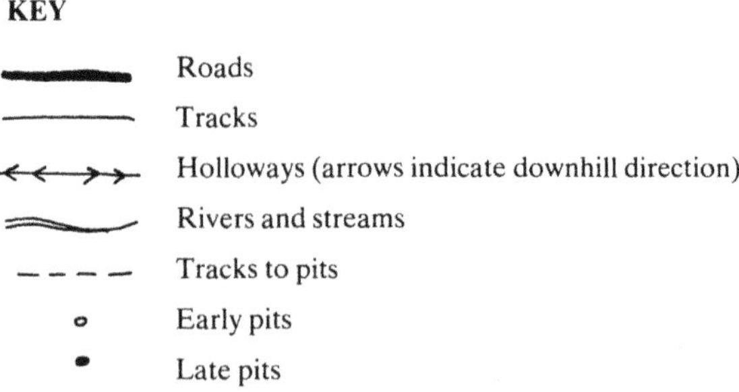

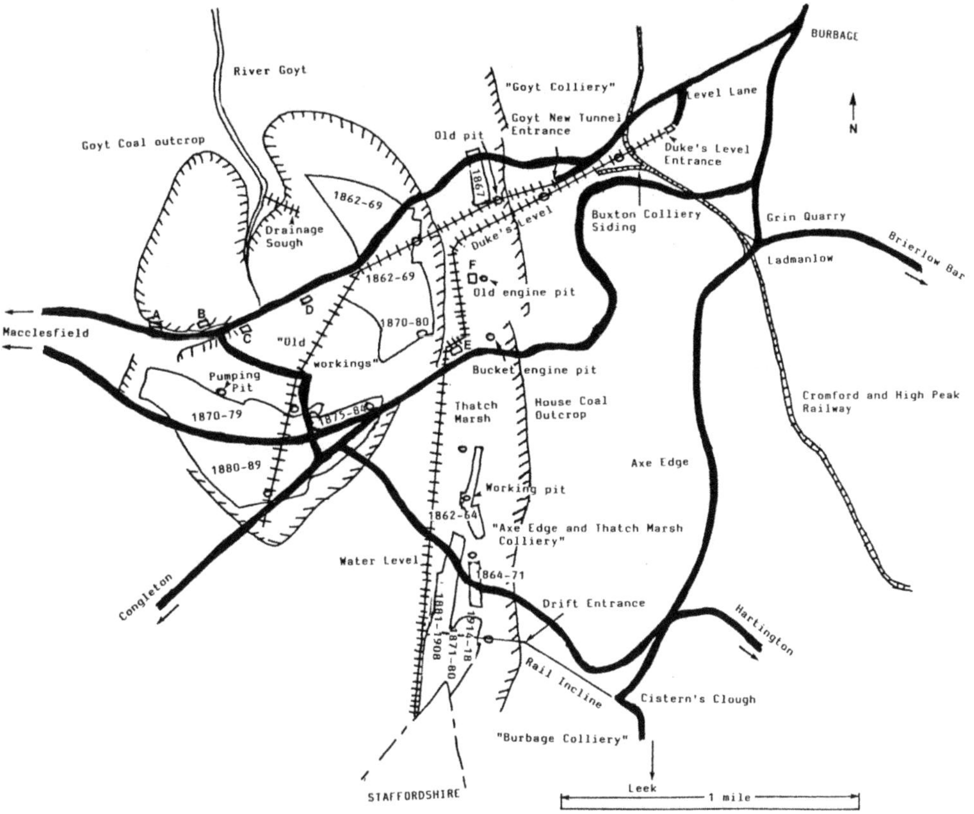

Fig. 3B. Later coal workings and road/rail network c. 1831

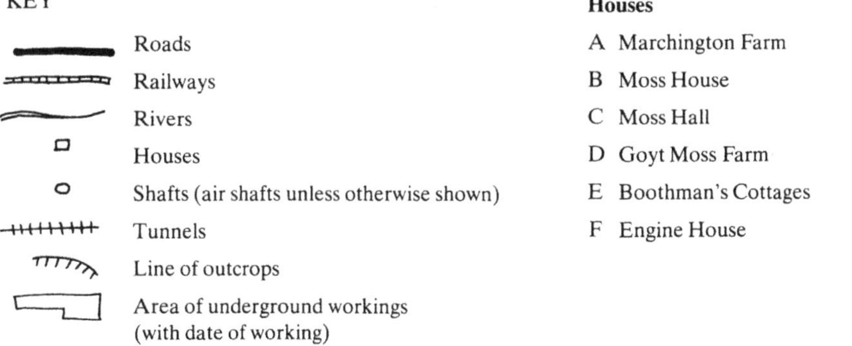

continues and rejoins the line of the turnpike. Slightly west of this point the early route is plainly visible as holloway sections a short distance to the north, and parallel with, the turnpike. At the line of the outcrop however the holloway diverges to follow different routes leading down to the north east corner of the Goyt coal workings.

The fourth holloway can be traced from near Dalehead, on the minor road to the east of Axe Edge running towards Earl Sterndale. Again, the best evidence for the route begins at the boundary of the limestone. A holloway is apparent on the southern side of the minor road for about two hundred yards west of Dale Head; its route then crosses the present road and runs north of it for a short distance before merging with it. A few hundred yards east of the A53 the holloway reappears as a separate route and heads in a more or less straight line towards Axe Edge. However, on reaching the A53, which started life as a turnpike built in 1765, the natural ground level is at least six feet below the level of the road surface of the A53. It is apparent that the holloway existed well before 1765 and the turnpike was built across it with a substantial supporting wall on its downhill side. Faced by this obstacle, users of the holloway created a second branch which veered off the original route in a south westerly direction in order to make a route by which the turnpike could be crossed.

The same features are apparent on the uphill side of the A53. The original holloway heads straight up the hill, up to 30 feet wide at the top and 10-12 feet deep, while the post-1765 diversion rejoins it about 250 yards from the A53.

The holloway heads up the hill and crosses a shoulder of Axe Edge at a height of about 1730 feet on to a flat and marshy plateau. Here there is no holloway but the route can be traced towards a distinctive outcrop of rocks about four hundred yards away because a crude causeway was made across the marshy area from stones, mainly about 1 x 2 feet, which are now covered by several inches of vegetation. The causeway would have assisted the heavily laden packhorses when crossing the marsh. The coal seam outcrops shortly past the rock outcrop.

The continuation of this holloway can be traced on the map from the Hartington Enclosure Award 1804[2]. On this map the Dale Head road is referred to as the 'Coal Road' and this description is used for a route passing from Dale Head past Harley, High Needham and Madge Dale[3] to Hartington. This route has the great advantage of being along a ridge with relatively few gradients until the final descent to the river valley level at Hartington. It seems likely that this route was in use for coal traffic prior to 1804 - indeed the Enclosure Award refers to it in the following terms:

'Coal road: which said last described road so far as the same goes over ancient inclosures we have set out and appropriated in lieu of the old road over the said inclosures leading from Dale Road to High Needham which is from henceforth to be stopped up and discontinued except only as to such small parts thereof along which the Coal Road shall happen to pass.'

The development of the turnpike system around Buxton had a profound influence in improving the accessibility of the coal field and allowing horse drawn carts to be used in place of packhorses. The first Buxton to Macclesfield turnpike was built in 1759 and it is likely that this prompted the major exploitation of the coal seams via underground workings. The Buxton-Leek turnpike followed in 1765, the Brierlow Bar-Ladmanlow turnpike in 1773 and the Buxton-Congleton turnpike in 1789 so that within 30 years a system of turnpike roads had developed in the vicinity of the coal field. The major development subsequent to that time was the construction of the revised route for the Buxton-Macclesfield turnpike in 1821.

In addition to this rapidly developed network of longer distance routes, many minor routes were developed to link the coal field workings with the turnpike system. A stone surfaced route approximately six feet wide and quite adequate for carts runs from the original Buxton-Macclesfield turnpike southwards, more or less parallel to the outcrop of the House Coal seam, towards the Dane Valley. Near the fourth holloway, described above, two additional routes were constructed. One of these, presumably the earlier one (Fig.3A), is a deliberately made route linking the outcrop area with the Buxton-Leek turnpike at Cistern's Clough, which would have served as a cart route for coal transport between 1765 and 1804. It is marked as a footpath on the Ordnance Survey map. The other was constructed following the 1804 Hartington Enclosure Award (Fig. 3B). It is better graded than the earlier one and links up directly with the 'coal road' to Hartington described above; it is also referred to as the 'coal road' as far as the junction with the Congleton turnpike. The latter route is the present minor road leaving the A53 to the west near Cistern's Clough.

In addition to those roads or tracks there are several stone surfaced tracks from the turnpikes into the area where the Goyt Seam was worked from the surface, which are described in more detail in Chapter 5.

A significant development in the communication system was the construction of the Cromford and High Peak Railway, which opened in 1831. The original plan was to link the midlands area around Derby and Nottingham with the Manchester region by a canal, linking the Cromford Canal at Cromford with the Peak Forest canal at Whaley Bridge. In the event, for cost reasons, a railway was constructed through the Peak District from Cromford to Whaley Bridge with nine inclines where trucks were wound up or down by stationary engines, the trucks being drawn along the sections between the inclines originally by horse and later by steam locomotives. The CHPR ran through Ladmanlow (see Figs. 2 and 3) and eventually it had branch lines to Grin Quarry at Ladmanlow, Harpur Hill Quarry and to the coal mines, as described in Chapter 7. It therefore provided a good means of transport of coal to the local quarries but at the same time it also reduced the transport costs of coal from more distant mines to the Buxton area. The competition of better quality coal from elsewhere therefore began to develop as the communication system improved. By 1863 main line rail routes had reached Buxton and the CHPR became leased to the LNWR network in 1861. Thereafter the Buxton area was easily accessible by rail.

CHAPTER FOUR

MARKETS FOR LOCAL COAL

The main market for Buxton coal was undoubtedly as a fuel in burning limestone for lime. For many centuries lime has found use in mortars for construction purposes and in agriculture to improve the quality of many types of land. The burning of limestone is easily achieved with more or less any fuel, and lime burning was a widespread practice in the limestone areas of the Peak District for a long time, with individual farmers burning lime for their own use.

When wood was freely available as a fuel, a farmer could quarry his own limestone and cut his own wood for this purpose but as supplies of firewood became scarce, probably during the 16th century, coal began to be more widely used as a fuel. Certainly by 1697, Celia Fiennes[1] could describe the country between Bakewell and Buxton as follows: 'you scarce see a tree and no hedges all over the country'. Thus the availability of wood was low by that time.

This trend probably began to establish the importance of Buxton as one of the main areas of limestone quarrying and limeburning in the country. In 1650 a survey to assess the property of the late King Charles I recorded fourteen limekilns at Dove Holes.[2] Pilkington[3] in 1789 records 'about' eight kilns at Grin, each with five hands burning 120 horse loads per day. As described earlier, being situated at the western boundary of the Peak District limestone country with a supply of coal available within one mile, Buxton was well placed not only to produce coal that could either be used locally or sent into the interior of the limestone plateau but also to send lime to areas to the west, which did not have access to alternative supplies of limestone. Thus a two way traffic grew up, with large quantities of Buxton lime moving into Cheshire and Staffordshire and Buxton coal moving towards Hartington and Taddington. In the 1850's coal was carted from the 'Level Pit' by George Buxton senior and junior. 'Billy' Robinson was another carrier of coal from the 'Level Pit' and he is depicted on a painting of 'Knox Rock' (Buxton) by R.O. Brunt[4]. Moses Longden was another very well known carrier of Burbage at this time.

During the nineteenth century farmers came from Monyash, Flagg, Taddington, Ashford and Wormhill to the 'Level Pit' to get coal, often in the early hours of the morning.

In 1734, Short[5], referring to limeburning, states ... with a coarse Coal got near, is burnt and carried into Cheshire and Lancashire and the neighbourhood, both for building and manuring the land'.

In 1780 the Duke of Devonshire[6] leased the coal rights on Goyt Moss to Robert Longdon of Countess Cliff Farm (near Harpur Hill), Richard Wheeldon of Cronkston Grange, Isaac Wheeldon of Buxton (an Inn holder) and Edmund Wheeldon of Buxton (an Inn holder); it is hard to identify the precise area covered by the lease from the description in the lease; the northern boundary was the 1759 Buxton-Macclesfield Turnpike and the southern boundary was Tinkerspit Gutter. The eastern and western boundaries are described in terms only of fences and a staked boundary. The lease did not provide for any cash payments to the Duke of Devonshire. Instead the lessees agreed to supply to the Duke each year 'fourteen thousand horse loads of good and well-burned lime consisting with the ashes of 12 pecks to each horse load and carry out and set the same in an husbandlike manner in proper heaps for spreading at the expenses, costs and charges of ' ... the lessees ...' upon such part of the lands and grounds (of the Duke) within the Parish of Hartington ... not further distant from any of the coalmines ... than a place called Cronkston Grange[7] in such places as (the Duke) assigns. The limekilns for burning such lime from time to time to be erected and made at the cost (of the Duke) and the places for erecting such limekilns for burning such lime and for getting stone to be burned into such lime to be set out and fixed (by the Duke, within the distance of Cronkston Grange, as before)'.

On the face of it, the Duke appears to have secured a good bargain; no details are given of the state of the mine workings leased to the lessees, but it seems likely that only the Goyt Seam was involved.

The late eighteenth century was a time of great interest in agricultural improvement and increasing the quality of land by drainage, fertiliser use or liming was very much in fashion. The Duke's lease therefore provided a means of improving the quality of his land at the expense of the lessees and left the lessees with any coal surplus to the amount required to burn lime for the Duke.

Farey[8] records that in 1783 the Duke of Devonshire caused a considerable tract of heath covered land 'to be improved under the direction of his Agents, Mr Robert Longsdon and Mr George Brassington, by spreading 1500 bushels of lime per acre on Hind Low, Sticker Hill and others, near to Hill Head Farm, the charge amounting to 2d per bushel including loading and spreading: some doubts have, however, been expressed of the propriety of some parts of this charge to His Grace: the effect, however, though slow, was striking; the heath being exterminated by the lime, a sweet and good herbage has succeeded ...'

Taken together, these two accounts suggest that the 1780 lease conditions did not remain in operation for very long; by 1783, Robert Longsdon was working with George Brassington rather than the Wheeldons and the Duke was paying 2d a bushel for the lime rather than receiving it free in recognition of the lease of the coal rights. Nevertheless, as the prevailing price for lime was about 8d per bushel at the time, he was receiving it at a below average price. At 4 pecks to a bushel, the lease provided for 42,000 bushels of lime per year to be applied to the Duke's fields which, at the high application rate of 1500 bushels per acre, would have improved 28 acres per year.

Farey[9] makes several other observations concerning the use of Buxton coal for limeburning. At Buxton (Grin Low) he notes that 'their coals (from Thatch Marsh and Goyte Moss) being bad, slaty and brassy, occasioned their Lime Ashes to be so heavy that the distant farmers from Cheshire etc. would not carry them away but, if measured with the Lime, would pick them out and throw them on the ash-heaps, and in such a way they had indeed been in part accumulated'. The large ash-heaps are still visible in Grin Low Wood and the heavy lumps of coal residue are still apparent.

At Blackwell, near Taddington, Mr Joshua Lingard[10] used Pye Kilns (i.e. large stacks of alternate layers of stone and coal covered with turves and fired at the bottom) rather than the normal small running Kiln (a stone walled kiln, kept hot continuously with stone added at the top and lime withdrawn at the bottom); he found that a running Kiln required one bushel of small Thatch Marsh coals to make two bushels and a half of lime whereas 'in a large Lime pye of 6000 bushels one bushel of the same Coals will make three to four bushels and a half of Lime'.

At Newhaven 'Mr Timothy Greenwood[11] uses a great deal of Lime and burns it in Pye-kilns or Pudding-pyes as some call them'. The method of construction of the Pye is described 'and the Pye is left to burn for five days, if good Coals from the Wharf at Cromford are used, or ten days if the Thatch Marsh Coals are used'.

Taken together, these extracts show the use of Thatch Marsh and Goyte Moss coal, despite an inferior quality, for limeburning as far as Hartington and Taddington; this information agrees well with the evidence on roads given in Chapter 3. The construction of the Cromford Canal to Cromford enabled better quality coal from the major Derbyshire coalfields to establish a market as far as Newhaven, while coals from Staffordshire, the Ringinglow area and Whaley Bridge area would have decided the marketability of Buxton coal in other directions. There would have been some market for coal west from Buxton but coal from the collieries at Poynton, Macclesfield and the Dane Valley would have advantages there.

In addition to limestone, another great natural asset of the Peak District was lead ore; the smelting of ore to make lead was a small scale operation involving local supplies of fuel until cupolas were introduced in the mid 18th century and the smelting of lead ore began to concentrate at relatively large sites conveniently situated for good supplies of fuel. Up to that time it is therefore possible that Buxton coal was used in the smelting of lead near to the mines in the Hartington, Taddington, Monyash and Chelmorton areas. This use would have declined as the larger cupolas were built on the eastern side of the Peak District.

There would also have been a domestic market for coal, evidenced by the fact that one of the seams was called the House Coal Seam. In 1801, the population in the market area was only about 4000, so the domestic market would not have been large. The population of Buxton grew steadily during the 19th century but, as rail communications improved from 1860 onwards, better quality coals from Lancashire etc. would have offered strong competition. However, coal began to achieve dominance over wood as a domestic fuel during the 16th century so it is likely that throughout the 17th and 18th centuries local coal served to warm Buxton's most prestigious buildings - The Old Hall, built c.1570 and the buildings, mainly hotels, in the Crescent, built c. 1780. Indeed, it seems likely that the 5th Duke of Devonshire, having seen the communications to Buxton improve and having constructed the Crescent, would have encouraged the development of a local supply of coal to ease the chill of Buxton evenings during the Spa season.

CHAPTER FIVE

EARLY DEVELOPMENT OF THE COALFIELD

Before the introduction of steam power, coal mining technology developed slowly. Fig.4 illustrates five typical stages in this development in a hilly area with a single coal seam dipping at a fairly low gradient.

The divisions between the various stages of mining are not hard and fast nor is the chronology rigid. Development occurred at different rates in different parts of the country - stage five workings are known as early as 1407 in Durham and stage one working is in operation today on a grand scale in the form of opencast or strip mining.

The type of working practised in any particular situation depended on geological conditions and economics. The main technical problems of early mining were controlling falls of ground and flooding. Going through the various stages of development required better means for dealing with these problems, as seams were worked to greater depths. This required greater capital investment which in turn required workings to become larger to repay the extra investment. Large workings lead to ventilation problems and eventually to the explosion problems that beset the industry during the 19th and 20th centuries. Thus, in stage three workings for example, the deeper the shaft the larger the workings had to be to pay for the cost of the shaft sinking, and the better supported they had to be to survive for the longer period of extraction.

In all stages considered here, extraction of the coal was by pick and shovel techniques and the main source of energy available for moving coal and waste out of the mine and for pumping and for other purposes was man or horsepower until the introduction of steam power, initially from 1712 for pumping and later for other purposes. In a few locations water power was used for pumping.

As far as the early development of the local coalfield is concerned documentary evidence is sparse.

Galloway[1] records that 'the sea coal in the Forest of Macclesfield was committed to the charge of a forester appointed in 1382'. Nef[2] states, with reference to the Cheshire Coal Field in the 17th century, that there is little evidence of any marked development in the output of coal in this district'.

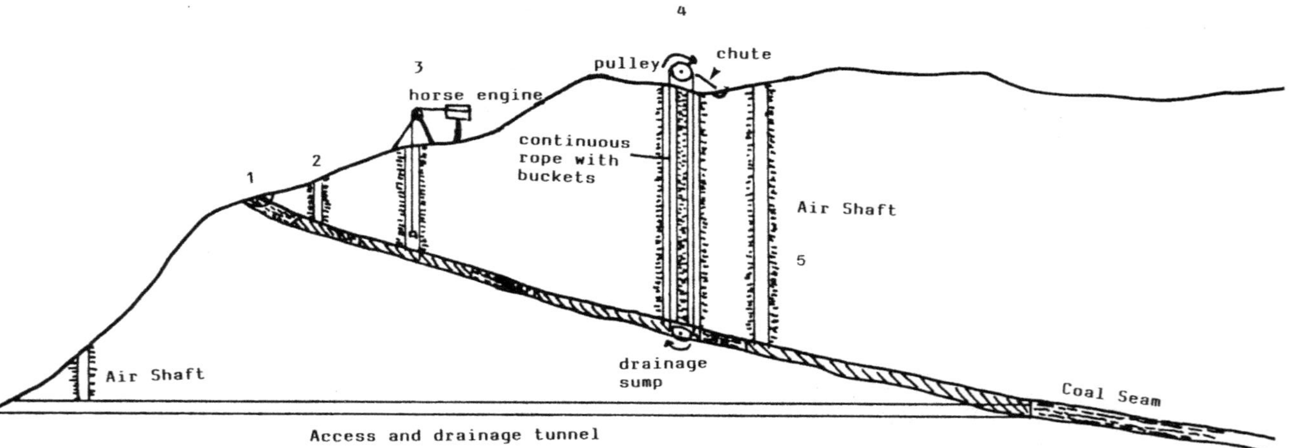

Fig. 4. Stages in the development of mining layouts.

KEY

1	Pit on outcrop
2	Early pit with shaft
3	Later pit with deeper shaft, worked by horse engine
4	Deeper shaft with bucket engine for drainage
5	Drift mine, access and drainage via tunnel, ventilation via air shafts

He records that Parliamentary Commisioners who surveyed the manor and borough of Macclesfield in 1652 did not find any improvement in the value of the tree 'pitts or delfs' which their predecessors had surveyed in 1611. Nef also refers to the dispute before the Star Chamber in 1606 concerning a mine at Fernilee in Derbyshire, previously described in Chapter 2. He concludes that there was 'no considerable demand for coal in the hilly sparsely settled district along the Cheshire-Derbyshire county line'.

View of typical, small, late eighteenth century coal mine.

Early evidence of coal mining activities nearer to Buxton comes from contemporary maps. A map[3] of the area around Axe Edge, Dane Head and Dove Head, with an estimated date of 1600, illustrates lands in dispute between John Claye and William Gilberte - 'pcell of Wharnford pasture wch Hartington mene make clayme unto'. The map shows a quarry and coal pits in this area.Interestingly, a similar dispute is recorded during the surveying for the Hartington Enclosure Award (1804) as to where the boundary between Quarnford (Parish of Alstonefield, Staffordshire) and Hartington lay in this same area. Presumably the dispute had rumbled on for two hundred years; the value of the minerals in the disputed area may have given an extra edge to the argument.

John Ogilby's strip map[4] of the road from Derby to Warrington via Buxton and Manchester (1675) shows the road junction at Brierlow Bar 2 miles south of Buxton, turnings to Chelmorton and Longnor and a road labelled 'to the cole pitts' on the line of the present road (Burlow Road) from Brierlow Bar to Ladmanlow.

Jackson[5] states, without giving details, that it may be inferred from leases and conveyances that the local coal pits were being worked in the early seventeenth century.

The first comprehensive survey of Derbyshire was by Burdett in the period 1761-63 Burdett's map shows 3 coal pits located immediately to the north of the original (1759) Buxton to Macclesfield Turnpike and one to the south, near the 2 mile milestone (on the line of the House Coal seam).

Thus there is positive evidence to show that the line of the House Coal seam was known at the beginning of the 17th century and that exploitation of this seam was taking place during the 17th and 18th centuries. A 16th century start to the exploitation is a possibility.

Field evidence, aerial photography and various plans provide much evidence for early working and if interpreted correctly can assist in the identification of the different stages of mining.

Figs. 3a and 3b illustrate the workings. Fig. 3a shows stage one to three workings and the road network up to circa 1790. Fig. 3b shows stage four and five workings and the principal road and rail connections circa 1831.

The earliest method of exploiting a coal seam was to locate where the seam outcrops (i.e. breaks the surface) and simply to dig pits or trenches (sometimes known as grooves) into the exposed coal. The amount of coal that can be extracted in this way depends on the slope of the seam and, for a seam dipping at 1 in 6 like the House Coal Seam on Axe Edge, it was relatively small before the labour of shifting the overlying soil became excessive.

The next stage of exploitation was to sink narrow shafts into shallow seams and extract the coal underground in the manner described in Chapter 6. Without drainage channels or with only primitive means of removing water from the workings, the limits of such shafts, in terms of depth, would usually be determined by ground water considerations. The area extracted per shaft would be determined by seam depth and the cost of sinking shafts balanced against the extra labour of transporting coal underground for increasing distances.

Later pits are larger and further away from the outcrop. The sinking of a shaft down to the coal seam created a larger amount of waste which was usually formed into the circular track for a horse to be led around to drive a horse engine or 'whim gin'. The waste, consisting of earth and rock, was deposited on the uphill side of the shaft to give height to the overhead machinery and to provide a barrier against surface water draining into the shaft.

Horse engine or gin for raising coal.

It is likely that the House Coal seam was exploited before the Goyt seam because of its better quality, its proximity to the potential markets (Buxton and the limestone areas) and also because of a steady descent from the outcrop to the markets. (Laden packhorses from the Goyt seam would have had to climb 135 feet to reach the ridge before descending, and to travel ½ mile further to the point of use).

On the more or less straight run of the House Coal seam there are 130 pits in the Axe Edge/Thatch Marsh area lying in a band approximately 3000 yards long and 100 yards wide with an average separation of about 50 yards. The band stretches from the outcrop to a line above where the seam is approximately 50 feet deep, which presumably is the level set by local water and ground stability considerations.

In the early days of exploitation of the House Coal seam it is likely that only one pit was in use at a time, perhaps with only two or three men working on it during the summer months. At this rate, one pit might only produce four or five hundred tons per year of coal but would last four years or so. A quickening tempo of economic activity and more demand for coal as wood for fuel became scarce would have led to more intensive exploitation, leading to the introduction of horse engines and increases in the number of shafts in operation simultaneously. Certainly, some of the pits into the House Coal seam have visible remains of the tracks for horses operating the engines.

In the House Coal seam, one can conveniently divide the total area up into five blocks:

Berry Clough to 1759 Buxton-Macclesfield turnpike		
	11 pits in 40,000 sq.yd.	3,600 sq.yd/pit
1759 Buxton-Macclesfield turnpike to footpath		
	27 pits in 54,000 sq.yd.	2,000 sq.yd/pit
Footpath to 1821 turnpike		
	31 pits in 43,000 sq.yd.	1,400 sq.yd/pit
1821 turnpike to 'coal road'		
	38 pits in 152,000 sq.yd.	5,400 sq.yd/pit
Coalroad to Staffordshire border		
	23 pits in 71,000 sq.yd.	3,100 sq.yd/pit

Thus the intensity of extraction varied somewhat along the outcrop but the average of 2,770 sq.yd/pit for the whole area derived from 130 pits in 360,000 sq.yd. (75 acres), is reasonably representative and fairly similar to that for the Goyt seam.

Along much of the House Coal outcrop the ground rises to the west as the seam dips. This means that the seam depth increases quite rapidly and so stage three mining was not greatly utilised. The increasing depth of the shafts necessary to extract the coal permitted only one row of stage three pits, as shown on Fig. 3a, before deeper, stonelined shafts (which can still be seen) had to be constructed. Judging by Shorts[6] account this transition had taken place by 1734.

The distribution of the stage one to three pits on the Goyt coal seam of which there are approximately 40, is more interesting because of the complex shape of the outcrop. The earliest pits (stages one and two) were worked along the northern, north-western and eastern edges of the outcrop as shown on Fig. 3a. The remainder of the outcrop (which is displaced in several places by lateral faulting) does not seem to have been worked in this way, although the name 'Tinkerspit' on the southern edge of the outcrop may indicate some early mining activity.

There are 68 later (stage 3) pits on the Goyt Coal seam and most are laid out systematically with regular spacings. Because the deepest part of the Goyt coal seam is only about 25 yards below the surface, it would have been possible to exploit most of the seam by this method if required. The regular layout of the pits is well illustrated in Fig. 6.

Some of the earliest visible ground evidence consists of roadways. There is a straight road heading northwards from SK07 0217.7128, which is now under several inches of earth and vegetation. This road can be located by the different colours of vegetation along its route and it is clearly visible on aerial photographs. It averages 16 feet wide and is ½ mile long, terminating near the northern edge of the Goyt seam outcrop. It is crossed by the later 1759 Buxton-Macclesfield turnpike at SK07 0200.7167, the boundary walls of which are continuous and completely ignore the access roadway. Several spur roads can be identified about 25 yards long leading to the banks surrounding old shafts. It therefore seems likely that this is an area in which systematic extraction of the Goyt seam took placc prior to 1759. The length and separation of the spur roads implies that each shaft was located at the centre of a 50 yards square block of the coal seam giving it access to 2,500 square yards of seam. A cubic yard of coal weighs approximately one ton so that, in a four foot thick seam with an overall extraction rate of 75 per cent, each shaft would have access to about 2,500 tons of coal unless limited by faults etc.

The overall impression of the early stages of development in the House and Goyt seams from the available evidence indicates that the earliest workings (stages one and two) took place in the 17th and early 18th centuries. Stage three pits were developed in the 18th century and continued into the early 19th century. In the House seam these early stages of mining declined quite early in the 18th century because of the necessity of extracting coal at far deeper levels.

With the improvements in communications due to the expanding turnpike system, an increasing market for coal, and seams becoming deeper, more advanced stages of coal extraction were soon to be developed. The following chapter gives an in-depth account of the year 1790 which in microcosm straddles the transition, with conventional and more advanced stages of extraction operating side by side.

CHAPTER SIX

THOMAS WYLD'S ACCOUNTS

In considering the details of the coal mining operations in the Buxton coal mines, Thomas Wyld's accounts for the year 1790[1] form an invaluable starting point. As was shown in Chapter 5, there is documentary evidence and field evidence for coal mining operations at a substantially earlier date, but Thomas Wyld's accounts provide the first detailed insight into one year of operation.

These accounts consist of 26 fortnightly accounts for cash transactions at the Buxton coal mines for the year 1790. Thomas Wyld occupied a position, normal for that time, of managing the business on behalf of the owner (the Duke of Devonshire) and being responsible for all cash receipts from the sale of coal and all cash payments for labour, materials etc.

A complete fortnightly account is given in Appendix 2 for one of the periods in the sequence - the fortnight ending 12th March 1790 -and three general comments can be made on this account.

1. The accounts are not a complete financial account of the mining operations since certain major items do not appear in them. For instance, there is no record of any salary or wage paid to Thomas Wyld himself; items appear such as 'expenses of travelling to Macclesfield to buy powder' and 'expenses of drawing rope from Buxton' but there is no record of payments for the actual powder or rope; there are records for cash payments for small amounts of coal produced from the Goyt seam but there is no cash payment for the bulk of this production which presumably was taken direct to a local quarry and paid for on a monthly basis or similar.
2. Coal production and sales are recorded in terms of the 'corf' (plural: corves) or basket rather than in terms of weight and listed in two columns as 'scores' and 'corves'. The use of the corf as a coal mining measure was first recorded in 1539 and the use of scores (originally 21 corves but later 20 corves) appeared around 1665. The corf was simply a woven basket used for transporting coal underground and for winding it up the shaft; it was not a standard measure but varied in size considerably from one part of the country to another and from one century to another. Its capacity was generally in the range of 1-3 cwt

according to Nef.[2] In this discussion it will be assumed that the local corf approximated to 1 cwt, which has the advantage that 1 score corves would then approximate to 1 ton, although it should be remembered that this might be an appreciable underestimate.

3. The colliery accounts are subdivided into 3 different sections - for the Buckett Engine Pitt and the Rise Pitt (both of which were in the House Coal Seam) and the Goit Pitts (in the Goyt Coal Seam). The total system was referred to as the Buxton Collieries or Thatch Marsh and Goit Collieries. In analysing and discussing the accounts these 3 different sections will be considered separately because they were clearly at different stages of development.

   The accounts give insight into the nature of the mining operations in each section, the costs and profitability of the operations, and the size and nature of the workforce.

The Goit Pitts

The accounts give a clear picture of activities on the Goyt Coal Seam throughout 1790. The seam was being worked by a series of shafts about 20 yards deep into the coal seam. For each shaft,timbers would be set at the base of the shaft to stabilise the ground in that area and coal extraction would then take place by working outwards from the shaft in a roughly circular area.

The coal was broken and loaded into corves by pick and shovel techniques; the corves would then be carried or dragged to the shaft bottom. The underground operations would take place with only primitive ventilation and by the light of candles or simple oil lamps.

The workings were known as 'wallings' and 'stauls' out of which the coal was mined in blocks. Between the 'wallings' some ribs of coal were left, and passages called 'thurlings' were used both for ventilation and for removing coal.

The shaft diameter would probably be about 3 foot with the miners descending on ladders or on wooden pegs hammered into the shaft walls. The coal was wound up the shaft in the corves using a horse whim or gin. A typical horse powered engine is illustrated on p. 29. The horse was led around a circular track about 30 feet in diameter and was attached by a harness to a large pulley wheel. This wheel drove a smaller pulley mounted above the shaft which wound loads up and down the shaft.

Typically, a corf would be wound up the shaft by the horse engine, swung to one side on reaching the surface and its contents would be discharged into a waiting cart or on to a small stockpile.

Early plateway as used in the Goyt tunnel.

The limitations on this type of mining were set by ground stability (with the ever present risk of shaft or tunnel collapse), ground water (with the ever present risk of flooding and the inability to mine below the prevailing water table unless some form of drainage measure was introduced) and difficulties of transporting the coal to the shaft which often meant a person on hands and knees dragging the baskets along the floor or carrying it on his (or possibly her) back bent double in the confined tunnel. For these reasons in shallow seams each pit only tended to work a relatively short distance out from the shaft bottom, it being cheaper to sink a new shaft into the seam than to extend the underground tunnel further. Water was removed initially by buckets, wound by hand or by horse gin, and later by the use of primitive 'rag and chain' pumps.

This type of mining operation was common in Britain during the 17th and 18th centuries but in the second half of the 18th century it was being progressvely superceded by deep mining techniques and by the introduction of steam power for pumping and shaft winding.

The mining operations can be considered in two broad categories - coal production and development and maintenance work:

Coal Production

At the begining of the year, 3 shafts were in operation designated Goit No. 1, Goit No. 3 and Goit No. 4. For each shaft, the output is recorded and the name of the team leader; the team was paid 10d per score corves and the leader would have had the responsibility for sharing this money out with the team members.

Output details are summarised in Appendix 3. Fortnightly output varied quite widely between 51 score 11 corves and 214 score 1 corf per shaft, but a value of about 120 score per shaft is fairly typical. This would have resulted in a £5 payment to the team for 12 days work. Day rates (paid mainly on maintenance/development work) went up to 2/2d per day and it is likely that the production workers would have expected at least as much income as the best paid day rate workers, i.e. 26/- per fortnight; this suggests that a typical size of team engaged on coal production was 4.

It can be seen that Goit No. 1 ceased production after the fortnight ending 12/2/90 but 3 shafts were kept in production by starting work on Goit No.5. Goit No.3 and No.4 ceased production for 10 weeks thereafter. Goit No.2 resumed production for 3 fortnightly periods (presumably this shaft had been in production in 1789 and temporarily abandoned); Goit No.4 resumed production when Goit No.2 ceased but only for 5 fortnightly periods and Goit No.3 resumed in the fortnight ending 18/6/90 but only for 7 fortnightly periods.

Meanwhile Goit No.5 ceased production in the fortnight ending 21/5/90 and Goit No.6 came into production in the following fortnight, continuing in production until the end of the year. Goit No.7 and Goit No.8 came into production in the fortnights ending 16/7/90 and 30/7/90 respectively and both of these continued in production until the end of the year.

Thus of the 26 fortnightly periods in the year there were 5 periods in which 2 shafts were in production, 16 periods in which 3 shafts were in production and 5 periods in which 4 shafts were in production.

Production details for 1790 were

Shaft	Periods in production	Output (score - corves)
Goit No. 1	3	428 - 9
Goit No. 2	3	330 - 3
Goit No. 3	13	1339 - 0
Goit No. 4	11	1326 - 17
Goit No. 5	7	1136 - 14
Goit No. 6	16	2385 - 19
Goit No. 7	13	1525 - 1
Goit No. 8	12	1494 - 18
	TOTAL	9967 - 1

Taking the corf as equivalent to 1 cwt, this represents, in round terms, 10,000 tons of coal production. In the peak period of production (fortnight ending 10/9/90) 623 score 19 corves were produced. There is no evidence of seasonal fluctuations in production and operations continued at a similiar rate during the winter and the summer, although the surface operations were taking place in very exposed conditions. Production was down somewhat in the final fortnight of the year - possibly the workforce had one day off for Christmas.

The associated operations of shaft winding and banking are also detailed in the accounts. Normally, coal was wound from the Goit pits by means of horse engines but when Goit No.8 came into production in the fortnight ending 30/7/90 it clearly did not use a horse engine; instead an entry appears 'Thomas Goodwin and partner (or James Sutton and partner) winding coal'. This job was paid at 3d per score and would have involved the two men operating a simple windlass arrangement. From 30/7/90 to the year end there was always one less horse engine in use than shafts in production.

It is worth considering the details of the winding operations. In one 12 day working fortnight, probably with a 12 hour working day, James Sutton and partner wound 200 score corves, i.e. 4000 corves in 144 hours or roughly 23 corves per hour. This is equivalent to one wind every 2 minutes throughout the working day. The sheer monotony and drudgery of such a task is hard to appreciate. Similarly, each horse in the horse winding engine was led by a lad of age 10-12 for 12 hours per day, while similar amounts of coal were wound.

One interesting insight into the reliability of the corf as a standard measure stems from the fact that more corves were wound out of Goit No.8 than the team were paid to produce. In the fortnight ending 22/10/90, for example, 108 score 16 corves were wound out but the team was only paid for 103-8. Over the 12 periods covered the total figures were 1566-2 and 1494-18 respectively. Since each corf would arrive at the shaft bottom and be wound straight up it is not obvious how this discrepancy arises; it may be due to winding corves of dirt in addition to coal or it may be faulty record keeping.

A third possibility is that the 'score' used underground consisted of 21 corves and the 'score' used on the surface consisted of 20 corves. It was noted earlier that the score was originally 21 and then became 20. This would mean that 2176 corves were wound out and the team paid for 2171 corves in the former example and 31,322 corves wound out and the team paid for 31,392 corves in the later example. This relatively close agreement supports this third possibility.

The banksmen would take the corf from the winding rope when it reached the top of the shaft and replace it with an empty one for returning to the workings. The full corf would either be emptied into a waiting cart or emptied on to a stockpile. Again, such work would be arduous and repetitive. During the period up until 26/3/90 and from 28/8/90 to the year end there was one more banksman working than there were shafts in production. The spare banksman probably kept the records of production at each shaft and collected cash for cash sales. During the middle period of the year there were 2, or on two occasions, 3 banksmen in excess of the number of production shafts. This coincides with a period of fairly busy development work.

At the beginning of September a new task appeared in the accounts and remained there until the end of the year:

'Edward Ward scutching the coal 12 days at 1/-'

Scutching is a word meaning trimming or shaping masonry with a hammer, and in this context would mean breaking the coal lumps to a more manageable size, possibly to supply additional coal to the domestic market.

Development and maintenance work

The development and maintenance operations are described in a number of places:

Fortnight ending 21 May 1790

Thos. Heald	sinking in shale 22¼ yd at 4/- per yd	4. 9.0
,, ,,	,, ,, rock 11 yd at 15/- per yd	8. 5.0
,, ,,	,, ,, bottom coal	1. 1.0
,, ,,	making a new road 50 yd at 1/- per yd	2.10.0
,, ,,	making a new sough	1.6
,, ,,	putting down Air Trunks	1.6
,, ,,	setting up head gear	3.0
,, ,,	making gutter round the pitt	1.6
,, ,,	setting 12 pairs of timbers at 1/-	12.0
		17. 4.6.

In the following fortnight (to 4 June)

J. Hudson	to shifting the engine and making a new mount	1.15.10
Isaac Lomas	driving endway to loose the new pitt 4 days at 1/6	6.0
William Warren	,, ,, ,, ,, ,, ,, ,, 3 days at 1/6	4.6
Joseph Tamblin	,, ,, ,, ,, ,, ,, ,, 5 days at 1/6	7.6
Edward Turner	driving endway to loose the new pitt 1 day at 1/6	1.6
Samuel Turner	,, ,, ,, ,, ,, ,, ,, 12 days at 1/6	18.0
John Brown	,, ,, ,, ,, ,, ,, ,, 4 days at 1/6	6.0
Thos. Ashmore	,, ,, ,, ,, ,, ,, ,, 5 days at 1/6	7.6
Samuel Ward	winding at the new pitt 2 days at 1/8	3.4
		4.10.2
John Brocklest	sinking in shale 14 yds at 4/- per yd	2.16.0
	,, ,, rock 5½ yds at 15/- per yd	4. 2.6
	setting 7 pairs of timber at 1/-	7.0
	setting up the gears and gutter 2 days at 1/6	3.0
	making a new road 25 yds at 1/-	1. 5.0
		8.13.6
Thos Sutton and one other, sinking 32 days at 2/-		3. 4.0
Geo Yates and one other, ,, 32 days at 1/8		2.13.4
		5.17.4

In the following fortnight (to 18 June)

Jm Nadin	sinking in shale 12 yd at 4/6 per yd	2.14.0
,,	,, ,, rock 2 yd at 15/- per yd	1.10.0
,,	getting and setting up head gear 2 days at 1/6	3.0
,,	setting 5 pairs of timber at 1/-	5.0
,,	making 20 yd of new road at 1/- per yd	1. 0.0
,,	sinking of 2 yds in the top rock at 5/6	11.0
Wm Trafford	driving endway 2 days at 1/2	2.4
Wm Nadin	,, ,, 2 days at 1/2	2.4
Wm Warren	getting stone 1 day at 1/6	1.6
Thos Goodwin	winding over the sinkers 22½ days at 1/-	1. 2.6
,,	breaking stone on the road 3 days at 1/6	4.6
Thos Sutton and partner	sinking 23 days at 2/-	2. 6.0
Geo Yates ,, ,,	,, 24 days at 1/8	2. 0.0
Jm Barker ,, ,,	,, 19 days at 2/-	1.18.0

In the following fortnight (to 2nd July)

Isaac Greenough	sinking in shale 2 days at 2/-	4.0
,, ,,	driving an endway in the coal 15 days at 2/2	1.12.6
,, ,,	setting the shaft bottom 1 day at 2/-	2.0
,, ,,	getting stone 3 days at 2/-	6.0
Geo Yates and partner	getting stone 17 days at 1/8	1. 8.4
Francis Ozbeldestone	116 yd of joist at 1½d	14.6
	427 yd of railing at 1d	1.15.7
	2 gates at 1/8	3.4
	making planks 2 days at 1/8	3.4
	sawing timber 2 days at 1/6	3.0
		6.12.7

In the following fortnight (to 16th July)

Jn Barker	sinking part of No. 8 13¼ yd at 4/-	2.13.0
,,	,, in rock 1 yd at 15/-	1. 6.3
,,	,, in bottom coal 2510 c at 2/-	5.0
,,	making a new road 8 yd at 1/-	8.0
,,	driving endway 10 yd at 2/-	1. 0.0
		5.12.3
Samuel Ward	sinking No. 9 in shale 21 yd at 4/-	4. 7.0
,,	rock 4½ yd at 15/-	3. 7.6
,,	setting up headgear, putting in trunks 3 days at 1/6	4.6
,,	setting up 4 pairs of timber at 1/-	4.0
,,	making a new road 20yd at 1/-	1. 0.0
,,	sinking in bottom coal 2s10 c at 2/-	5.0
,,	driving endway 1 yd at 2/-	2.0
		9.10.0

Fortnight to 27th August

James Hudson	sending up shale to enlarge the banks 4 days at 1/2	4.8
Thos Head	making 62 yd of new road at 1/-	3. 2.0
,,	repairing roads 5 days at 2/6	12.6
Jn Dixon	driving endway to loose the new pitts 61 days at 1/8	5. 1.8

Geo Yates	getting stone 1 day at 1/6	1.6
Jn Wilde	filling an old pitt	5.0
Jn Bradley	getting stone 12 days at 1/6	18.0

Fortnight to 24 September

J Sutton	repairing the road 1 day at 1/6	1.6
J Bennett	getting stone 12 days at 1/6	18.0

Fortnight to 5 November

J Bennett	getting stone 12 days at 1/6	18.0
Gilbert Plant and partners	guttering 40 yd at 1/2	2. 6.8

Fortnight to 3rd December

Samuel Ward	raising the mount and headgear 21 days at 1/2	1. 4.6

Fortnight to 17 December

Jn Wheeldon & 6 others	loading shale 7 days at 1/2	8.2
son	loading shale 1 day at 0/6	6
Josh Brocklest & Co	loading shale 7 days at 1/2	8.2
Wm Norton	getting stone 11 days at 1/6	16.6
J Plant	mending roads 1 day at 1/6	1.6

The accounts identify a complete cycle of development and maintenance work from sinking new shafts and making access roads, through repairs of existing shafts and roads to the filling in of old pits.

From the accounts it is clear that Goit Nos. 1-5 were in production before any of the 1790 shaft sinking operations and were therefore sunk in 1789. Four additional shafts were sunk in 1790, Goit Nos. 6-9 of which Goit Nos. 6-8 were rapidly brought into production but Goit No.9 did not produce in 1790. This shaft sinking work was concentrated into the months of June, July and August, presumably because the ground would have been much drier at that time. The shaft sinking operation would have been a dangerous one with a risk of ground collapse and flooding since the area of Goit's Moss where the shafts were sunk is, as its name implies, a marshy/peaty one. The spoil from the shafts was used to build a circular bank about 30 feet diameter on the uphill side of the shaft mouth which served both as a track for the horse operating the engine and as a small dyke helping to keep water away from the area around the shaft mouth. Thus there would be an advantage in sinking shafts during dry conditions and relying on the bank to keep water under control during wetter times.

The depths of the shafts varied but about 20 yds of shaft sinking in shale and about 5 yds of shaft sinking in rock is fairly typical. Setting up the headgear (a wooden framework for the shaft winding operation) generally took 2 days and setting timber supports at the shaft bottom is also mentioned. The only mentions of ventilation of the workings are the reference to 'putting down Air Trunks' which were wooden ducts made from planking, which helped to direct what little air movement was achieved by wind effects into the inner regions of the workings.

The access roads to the shafts were made by the development teams from stone quarried locally. The length of road associated with each new shaft varied; one entry is as little as 8 yd and one is as high as 62 yd. In total, 185 yd of road were constructed during the year. The entries are broadly consistent with a development pattern of a central road with spurs about 25 yds long off to shafts on either side, the spurs being at about 50 yd intervals.

A shaft sinking programme in May-July eased problems by making three shafts available for immediate production with a fourth in reserve.

The Buckett Engine Pitt

This pit can be located because the name is still used on the Abandonment Plans. The shaft is located at SK07 0283.7140, approximately 20 yards north of the present Buxton-Macclesfield Road and visible from it (see Fig.3b). The present visible remains of the Buckett Engine Pit consist of a circular track 45 feet in diameter on a raised bank about 5 feet above the prevailing ground level. The tops of two stone lined shafts, about 4 feet 6 inches diameter and 9 feet 3 inches apart, are still visible above the sealed off level.

The term 'Buckett Engine' describes a device somewhat similar to the horse engine described previously except that buckets of leather or wood would be mounted at intervals on a continuous rope passing over a pulley at the top and at the bottom of the shaft. This engine would be horse driven and would normally operate continuously, giving a steady flow of full buckets to the shaft top with the empty buckets returning similarly. Bucket engines were used both for winding water for drainage purposes and for winding coal; in the former case, the water would discharge into an elevated drain channel as they tipped over the top pulley, while in the latter case the coal would discharge into a chute.

Despite its name, Thomas Wyld's accounts give no indication that the Buckett Engine Pitt was being used in this way. Unlike the Goit Pitts and the Rise Pitts, there is no reference to use of horses to power any engine and there is no indication that steam power was in use locally at that time. It therefore seems likely that the shaft, which would be about 130 yards deep, was sunk originally as an isolated part of the workings and was served by a horse powered bucket engine but was subsequently connected to another part of the workings and coal was moved out that way by 1790.

One of the two lined shafts of the Bucket Engine pit.

An alternative to this mode could be that as used in a 'Water Engine'[3] near Pott Shrigley where a weight of water descending one shaft raised a smaller weight of coal in the adjacent shaft. This principle required an underground drainage channel to remove surplus water.

What is clear from the accounts is that in 1790 coal was moved from the Buckett Engine Pitt to the surface along an underground canal or 'navygation'. There are frequent references in the accounts to activities related to the canal, of which the entries in Appendix 2 are fairly typical:

> Repairing the navygation
> Boating the coals
> assisting to boat coals
> repairing the boats by night

The position of this canal can be identified from the Abandonment Plans.[4] It began with a tunnel, known as the Duke's Level, driven from the end of the present Level Lane in Burbage (SK07 0370./7230) for a distance of 1,360 yards into the House Coal Seam. The mouth of this tunnel is at an altitude of 1170 feet AOD. An Ordnance Survey map of the district[5] shows a rectangular area of water at the tunnel entrance which presumably constituted a small dock (Fig.5).

On reaching the House Coal Seam the tunnel was then extended along the seam, at the same level, in a southerly direction. The water level is about 400 feet west of the bottom of the shafts of the Buckett Engine Pitt.

The Buckett Engine Pitt is 1600 feet south of the original Duke's Level, so presumably they were developed initially as separate entities; coal working in the House Coal Seam with an allied extension of the water level would subsequently have made it more logical to take the coal out horizontally in boats rather than vertically in buckets.

The Duke's Level was the major feature of the development of the underground workings, with a dual role as a drainage level and a transport system. It enabled the workings to be drained from the surface down to the 1,170 feet AOD level and greatly simplified the transport of coal out of the mine.

Similar developments took place elsewhere in the district at this time. The first such development in the use of boats for underground transport was by John Gilbert, the Duke of Bridgewater's canal/mining engineer at Worsley Colliery in 1759. According to Robey and Porter,[6] Gilbert was responsible for similar developments at Hillcarr Sough, started in 1766, Speedwell Level in 1774 and Ecton Mine in 1767. The Duke of Devonshire was directly responsible for the operation of Ecton Mine at that time and references to payments to Ecton miners appear in Wyld's accounts of 1790. It therefore seems quite likely that John Gilbert was also responsible for driving the Duke's Level for this coal mine and an estimate of about 1770 for the driving of this Level fits into the same pattern. Because of the costs involved in driving a tunnel, the Duke's Level was probably of quite a small cross section, about four feet across, perhaps. Coal was also boated out of the 'Blackclough' (or 'Beat') mine on the Derbyshire/Cheshire border.

Canal tunnel at Worsley; this arrangement was similar to that at the entrance to the Duke's Level.

There were at least two vertical airshafts from the Duke's Level to the surface at SK07 0843.7217 and SK07 0270.7093; the former of these is about 200 yards west of the large CHPR embankment near Burbage Reservoir. The shaft is now filled in but the spoil excavated from it still remains as a crude ring around the shaft mouth. Curiously, there is a very similar pile of rock only 30 yards away, suggesting that perhaps two shafts were sunk, one of which missed the tunnel. The second air shaft which can be seen to have been lined is 500 yards into the tunnel and is 130 yards deep. It too is capped and marked by an untidy heap of waste material, this time tipped down a slope.

The pattern of work was far more settled in the Buckett Engine Pitt than in the Goit Pitts. Throughout the year, production was in the hands of William Wild and Co. and, judging from production figures, the team was about 4-5 strong. With one exception (fortnight ending 17 December) the production figures were in the range 92-164 score corves as shown in Appendix 3. In the exceptional fortnight a good deal of maintenance work took place:

William Wild	driving navygation	10 yds at 1/-
William Wild	cleansing navygation	10 yds at 2/-
John Street + 5	cleansing navygation	60 days at 1/6
Joshua Brown	cleansing navygation	10 days at 1/2
John Plant	cleansing navygation	3 days at 1/6
Joshua Renshaw + 2	cleansing navygation	30 days at 8d
William Norton	cleansing navygation	1 day at 1/-

During this fortnight, 2 days were spent on normal coal production and the remainder on this exceptional bout of maintenance of the canal.

Production for the year was 3174 score 8 corves paid for at the rate of 1/4 per score.

Apart from the production team, the only regular underground jobs listed were 'boating the coals', i.e. transporting them by boat to the surface, and 'hooking and filling of coals'. The transport of the mined coal from the face to the boats is not listed separately and presumably was part of the production team's job. Two lads were paid 8d or 1/- per day for boating the coals, which would have meant wading in the water and towing the boat or legging it along for a distance of 3800 yards per journey. Daily production could be as high as 14 score corves or at least 14 tons; the shallow draught boats might have held a relatively small amount, perhaps as little as one ton, requiring 7 double journeys per lad per day. Since one would always need a boat waiting to be filled at the production point presumably a filled boat went out to be emptied and returned immediately, waiting for the second boat to be filled at the production end of the tunnel.

One man was paid 1/10 per day for hooking and filling of coals. This could refer to hooking the boats and filling them with coal.

The other regular daily jobs were at the surface where there were normally two men banking the coal and one man breaking the coals each day; these tasks paid at 1/9 and 1/4 respectively.

Apart from these daily jobs various maintenance and development jobs were noted.

Samuel Trafford & Co. had a regular job, which appears in each fortnightly account, of 'cleansing of the Lodge and Landing'; in some weeks '& Co' is replaced by '+ 4' and the payment was generally 8/2, implying about 5 man/days of effort. A lodge is a mining phrase for a building erected to protect surface machinery or installations, suggesting that there was a lodge at the landing stage for the canal, which would be the unloading point for the coal. The coal would be shovelled from the boats to the bank, unless transported in corves in the boat, and the need for regular cleansing suggests that shovelling on to a heap is more likely. There is still an accumulation of coal and waste material at this part of Level Lane.

Coal was burned at the bank and a record duly kept, so the Lodge, at least, may have been warm. One curious task was that of John Hibbert; in addition to his normal daily work of breaking coal, he received an additional payment, normally of about 2/6 a fortnight, for 'waiting over the men at nights'.

As a carpenter, John Renshaw was a key figure in the maintenance area and appeared to be the only craftsman employed. He worked full time for the Colliery as a carpenter (at 1/8 per day) but was paid additionally for certain tasks. 'Repairing the boats at nights', 'making timber at nights', 'repairing shafts at nights', 'making air trunks at night' were some of these. At times he was clearly overloaded.

He received a 5/- payment for working over hours in the fortnight ending 8th October and at the same time Thomas Kirk was being paid for 30 days at 6d per day for 'repairing of corves and making timber while the carpenter was making new boats'. Four weeks later Thomas Kirk was again mending corves, this time for 2 days at 9d per day. In the fortnight ending 19th November, William Bennett spent 7 days (at 1/6 per day) 'letting down a boat'. Whether this refers to one of John Renshaw's recently constructed new boats is not clear, nor is it clear what 'letting down' implies.

The canal gave problems throughout the year, culminating in the major maintenance described earlier. There are numerous references to 'cleansing the Navy and making dams', 'getting shale out of the Navy', 'getting a bar out of the Navy' and 'filling shale into boats out of the Navy'. On occasions there are references to 'turning water', which meant diverting water into the main drainage system.

The main development jobs during the year were

Charles Bagshaw	blasting in the rock	5 days at 2/-
John Street	blasting in the rock	3 days at 1/6
William Wild	driving an endway	27 yds at 6d
Thomas Heald	sinking of a trial pit	2 days at 1/8
William Wild	driving the Navy	5¼ yds at 1/- -
William Wild	driving the Navy	14 yds at 1/-
William Wild	driving the Navy	10 yds at 1/-

Thus the picture for the year is one of steady progress along an established pattern.

The Rise Pitts

There are no definite documentary records of which shafts correspond to the description used by Wyld of 'Rise Pitt', 'Old Rise Pitt' and 'New Rise Pitt'. It is clear that at the beginning of 1790, one pit was in operation known as the 'Rise Pitt'. Sinking of the 'New Rise Pitt' began in March (see Appendix 2) and when this came into operation in October 1790 the term 'Old Rise Pitt' was introduced to distinguish between the original and the new workings. It is also clear that these pits were in the House Coal Seam because the price paid for mining it and the selling price were identical with the Buckett Engine prices for House Coal Seam product. One further indication is that a total of 64¼ yards depth of shaft sinking is recorded for the New Rise Pitt in 1790.

The Abandonment Plans record a number of possible shafts in the House Coal Seam but, assuming that all of the New Rise Pitt was sunk in 1790 and that 64¼ yards is its true depth, all except two of these are too deep. The two remaining are designated 'old pit' and 'old Engine Pit' on the Plans and the depth of the former can be ascertained from a sectional elevation as 63⅓ yard. It is therefore assumed that the 'old pit' is the New Rise Pitt, because of the similarity in depth, and the 'old Engine Pit' is the Old Rise Pitt (these pits are shown in Fig. 3b) They are in reasonably close proximity to the Duke's Level and in a logical place for an early phase of the development.

The Old Rise Pitt worked intermittently throughout 1790 with 2 fortnightly periods of production in January, 4 in April/May and then continuous production from the fortnight ending 24th September to the year end.

Such breaks in production could have been caused by technical problems such as shaft collapse, encountering a fault in the coal seam, or excessive water accumulation, but there is no record of any remedial action taken to overcome such problems; during the non-productive periods there is no recorded activity in the Old Rise Pitt. The explanation for the breaks may be related to the availability of horses or horse engines. From 29th January to 24th September there were three horse engines in use, whereas both before and after that period four horse engines were in use.

Coal was wound out from the Old Rise Pitt using a horse engine, and from 29th January to 24th September, whenever the Old Rise Pitt was producing, there were only two horse engines in use on the Goit shafts. Conversely, when the Old Rise Pitt was not producing, there were three horse engines in use on the Goit shafts. This indicates that there were only three engines available for use between the Goit shafts and the Old Rise Pitt in that period. This problem was eased by using manpower for winding coal from Goit No. 8 when it came into production in July; and when the New Rise Pitt came into production in October it is apparent that a fourth horse engine had come back into operation.

Apart from its shaft winding problems the Old Rise Pitt did not have any unusual features. From its position relative to the outcrop it was probably about 120 yards deep. Its maintenance and development activities are only recorded as 'drawing shale', 'opening of the air road', 'repairs to shafts and road', 'ridding the endway' and 'setting timbers in the air road'. It produced 1067 score 4 corves of coal in 15 fortnightly periods.

The New Rise Pitt was a scene of more varied activity. Shaft sinking operations are first recorded in March and the shaft was completed in October at a cost of £23.7.10. The payment for shaft sinking in shale was initially 4/6d per yd compared to 4/- or 4/6d per yd for the Goit shafts, but this figure rose to 7/6d and then to 9/6d, presumably reflecting the increasing problems of sinking at depth.

In the fortnight ending 8th October, Charles Bagshaw and 5 others put in 20 days (at 1/6d) on 'shifting the engine, making the engine race and getting stone'. In the same fortnight is recorded 'by liquor given to the colliery workers at the opening of the new pitt 4/-'. Coal was also produced in this fortnight by Charles Bagshaw & Co. There was then a six week period during which further shaft sinking operations were paid for before regular production began in the fortnight ending 3rd December. Total production for the year was 200 score 14 corves.

Despite their proximity to the Duke's Level, the two Rise Pitts did not use the boat system for coal transport; they could, however, have used it for drainage as the shaft bottoms were at a higher level than the Duke's Level. Adjacent to the Rise Pitt entries, but separate from them, are two sets of references to a sough:

Fortnight ending 24th September:

Nick Kirk	cleansing the sough	12 days at 2/6
Sam Turner	{ cleansing the sough helping to load timber }	12 days at 1/2

fortnight ending 8th October

Nick Kirk	cleansing the sough	8 days at 2/6
Sam Turner	cleansing the sough	8 days at 1/2

This sough could have linked the Old Rise Pitt workings to the Duke's Level; the day rate of 2/6 is the highest recorded in the year and suggests a particularly dangerous and unpleasant job.

Financial Aspects

As described earlier, the accounts only related to cash transactions at the colliery but one can nevertheless get a fairly clear picture of the overall financial situation. Production and financial details for the year 1790, as taken from the accounts, are summarised in Appendices 4 and 5.

There are several noteworthy features about the cash sale figures both on their own and in conjunction with the production figures.

Firstly, the number of corves recorded as sold at the Buckett Engine and Rise Pitts exceeded the number recorded as produced by about 10% in each of the fortnightly records. Of the 45 records for these pits throughout the year the excess lay between 8% and 12% in all but 7 of the cases, with the excess in the remaining cases being as low as 3.4% or as high as 23.7%. This has two implications. It seems clear that all the coal produced in a fortnight at these pits was sold in that fortnight - there is no evidence at all of producing for stock or selling from stock. On the contrary, the evidence suggests an immediate sale of all coal produced - one could picture a queue of carts waiting to be filled at each pithead with the coal wound up the shaft either going straight into a cart or on to a small pile for breaking and then loading into a cart. It definitely resembles a 'sellers market'; in a sense the coal was underpriced if it sold so readily. A further implication is either that for production purposes a full corf was regarded as a 'heaped' corf, whereas for sale purposes it was regarded as a 'level' corf, so that the excess was literally skimmed off the top and sold separately, or that the corf for production purposes was 10% larger than the corf for sale purposes. Either way, the pattern is consistent: the number or corves sold exceeded the number produced by 10%. The consistency of the ratio suggests that the second explanation is the more likely, in which case the coal would have been unloaded from a 'production' corf on to a pile rather than directly into a cart.

Secondly, the banksman at the Buckett Engine Pitt (Samuel Bennett) may have had problems with arithmetic. The selling price of the coal was 4/2 per score, which means that 12 score would cost £2.10.0. Out of 26 sale entries for the year, 15 are exact multiples of 12, which made the fortnights reckoning much simpler, e.g. 156 score = £32.10.0. Possibly Samuel Bennett had a small cash float and rounded up or rounded down the fortnight's production figures to the nearest multiple of 12 whenever he could. In present times, working out the value of 139 score 17 corves at 4/2 per score can still present difficulties, even with a calculator to hand.

Thirdly, it is clear that most of the Goit seam production was not sold for cash. During the year, 9967 score 1 corf of Goit coal was produced (at 10d per score) whereas only 386 score 6 corves were sold (at 2/- per score). The most likely explanation for this difference is that 9580 score 15 corves of the Goit coal was sold either on credit or as an internal transaction to a user under the same ownership. The most likely outlet for this coal is the limeburning industry on Grin Low, only 2 miles from the colliery, or other major limeburning concerns, possibly under the control of the Duke of Devonshire. In any case, to get a complete picture of the profitability of the collieries one should credit the income side with the balance of the Goit coal production, at 2/- per score, i.e. £958.1.6.

On the income side in Table 1, the 'other' entry of £7.12.2½ consists of only two items:

supply of stone to Geo Slater	3. 9. 0
cash received from Geo Mellor and Wm Etches for kiln coal due in 1789	4. 3. 2½

Apart from confirming the use of the coals in kilns, the latter entry suggests that some credit arrangements existed.

Further insight into credit payments and debts comes from the expenditure side of the accounts. In the fortnight ending 29th January is reorded.

By Jn Ward's expenses collecting part of the Darbyshire debts	£1. 10. 0

John Ward was in charge of banking the coal at the Goit shafts and no doubt had well developed muscles; what his expenses were is not stated, nor is the amount of the 'Darbyshire debts' recovered recorded.

Other insight is given in the fortnight ending 19th November, when Thomas Wyld himself attended Macclesfield Fair and Chapel Fair 'when collecting the coal money'. His expenses (and those of his companions if any) were £3.15.2 at Macclesfield and 15/10 at Chapel. At Macclesfield Fair the breakdown of expenses was:

Liquor	£2. 3. 0
Eating	£0. 15. 0
Horses hay and corn	£0. 7. 0
Servants	£0. 7. 6
Unstated	£0. 2. 8
	£3. 15. 2

Considering that the amount spent on liquor for the colliery workers to mark the opening of the New Rise Pitt was only 4/-, the day out at Macclesfield Fair must have been a good party. Unfortunately it is not stated how many participated - possibly only Thomas Wyld himself from the colliery, established in a room in a Macclesfield pub entertaining a succession of credit customers coming to settle up their last year's account.

These entries give a glimpse of the wide ranging side of the business; they certainly suggest a network of credit customers for Buxton coal out to Macclesfield and Chapel receiving goods against a promise to pay at these annual fairs, which were important parts of the commercial activity of the region.

They also show something of the role of liquor in the transactions. On occasions liquor was used to recompense various workmen:

Liquor given to the Ecton miners and mason	14/-
Cash given for drink for the woodcutters	2/-

The reference to the Ecton miners and mason was made in connection with the sinking of the New Rise Pitt. The copper mines at Ecton, also operated by the Duke of Devonshire at that time, were much deeper and technically more demanding that the Buxton collieries. Presumably the Duke's agent borrowed the services of some of the Ecton miners and a mason for some specialised task, such as lining the shaft with stone.

On the expenditure side of the accounts, most of the outgoings relate to the direct costs of the mining operations, as described earlier. However, the 'others' entry is appreciable and covers purchases of materials, carriage on items, tolls, expenses etc.

The largest item of expenditure in 'others' was payment made to George Hawley, a blacksmith. During the year he supplied a wide variety of items including hammers, nails, sheaves, hooks, rakes, picks, wedges, plates, boat hooks, pulley wheels, corf handles etc. He also repaired many tools during the year. The total payments to George Hawley during the year were £80.13.6, which represents a sizeable proportion of the year's expenditure.

Timber was also purchased in various forms. Francis Ozbeldestone supplied 4100 ft of sawn boards at 1d/yd and, for a shaft sinking operation at the Goit seam, he was paid

116 yd. of joist		14. 6
427 yd. of railing at 1d		1.15. 7
2 gates at 1/8		3. 4
making planks	2 days at 1/8	3. 4
sawing timber	2 days at 1/6	3. 0
		2.19. 9

He also supplied boughs of trees at 2/6.

There are several references to payments for the carriage of timber, some of which are related to the sinking of the New Rise Pitt, but no accompanying references to payments for the timber.

On one occasion 'sixty oake topps for the use of the works' were purchased at £4.4.0, possibly large cross section timber for horse engine construction, and on another 'planks' at £2. John Wild supplied 2 cart loads of ash boughs at £2.2.0 and Sam Newton supplied 6 cords of wood at 2/- each.

Thomas Wyld incurred expenses of 4/- for going to Barleyford Wood for timber (+ 2/- for cash for the woodcutters as mentioned previously), 5/- on another occasion plus an assistant to Barleyford Wood and 4/6 for expenses to take up timber in a wood on a third occasion. There is a wooded area in the Dane Valley, near Rushton Spencer, at Barleyford Bridge which may have been his destination. On two occasions he went to Macclesfield to buy planks (expenses 2/6 per trip) and on other occasions to Macclesfield to buy powder (2/6), to Macclesfield for sundry articles (3/6) and to Kettleshulme to order corves and Macclesfield to buy tools (2/6). With the greater industrial development in Macclesfield as a textile town at that time, it appeared to be the natural source of supplies rather than Buxton, despite its much greater distance.

In fact Buxton only appears in the accounts incidentally:

drawing the new ropes from Buxton	1/4

The town and the collieries probably had little in common. Some payments were made to local Toll Bars - 11/2 to John Bennett at the Ladmanlow Bar on two occasions (Buxton/Leek and Brierlow/Ladmanlow Turnpikes) and 5/10 to Joshua Bennett at Gosling Bar (on the 1759 Buxton/Macclesfield Turnpike).

Other infrequent payments included the purchase of two horses at £12.12.0, one a black mare 5 years old and the other a bay mare bought of Joshua Francis of Buxton. Sometimes, horses were hired, at a cost of 1/- per day:

one horse drawing the engine at Goit	26 weeks at 6/-
Geo Slater and his horse leading stone to the roads	27 days at 2/6
Geo Slater himself	1 day at 1/6

One other significant item relates to a future development of the coal field. In the fortnight ending 16th July are entries

Jos Sutton making a platting .....[x]across the Goit River to the Cheshire Coal Pits	8 days at 2/-
Ed Greenough + 4 ,, ,, ,, ,,	30 days at 2/-
Jn Salt + 3 ,, ,, ,, ,,	33 days at 2/-
Jove Goodwin getting stone for above	27½ days at 1/6

[x]platting is a local word for 'bridge'.

As shown in Fig.3, the Goyt seam lies on both sides of the River Goyt, which at that time formed the Cheshire/Derbyshire boundary. Clearly, there were some pits active or in prospect on the Cheshire side with a natural outlet for the product alongside the Derbyshire developments. At one time there were 2 small bridges crossing the Goyt in this area and these may have been the ones described.

Taking all the above factors into consideration, a number of items should be added to the expenditure listed in Thomas Wyld's accounts to get a true picture of the profitability of the collieries. One is Wyld's income, which does not figure in the accounts. The top rate for day-wage jobs was 2/2 (excepting one obviously dangerous one) so that a salary of £50 p.a. would have given Wyld a clear margin above the manual workers.

Items such as purchase of ropes, timber, powder, tools, are indicated during the year, with no cash payments to cover them. As a rough guess, based on other costs in the accounts, £100 p.a. should have covered these with ease.

These two items have been added to the expenditure side in Table 1, together with the estimated income from additional sales of Goit coal to give an estimated profit for the year of £480.13.4.

It is hard to estimate the capital investment in the mines but it would appear to be fairly small. The major investment items were the driving of the Duke's Level, the sinking of the major shafts, purchase of boats, horse engines and horses. Most other items (and indeed some of the capital items such as shaft sinking) appear in the accounts. Guessing once more, £500 might cover past capital expenditure so that the estimated profit represents a very good return on the investment.

TABLE 1	**Production**	**Sales**	**Sales**	**Costs**
	S — C	S — C	£ s d	£ s d
Buckett Engine Pitt	3174. 8	3513.16	732. 0.10	448. 9. 2½
Old Rise Pitt	1067. 4	1168. 5	243. 7. 8½	96.14. 9
New Rise Pitt	200.14	220. 7	45.18. 1½	43. 1. 8
Goit	9967. 1	386. 6	38.12.7	599.13.0
Other			7.12. 2½	207. 1. 0
			1067.11. 5½	1394.19. 7½

INCOME		**EXPENDITURE**	
From above	1067.11. 5½	From above	1394.19.7½
Balance from sale		T. Wyld salary	50. 0. 0
of Goit coal at 2/-	958. 1. 6	Other purchases	100. 0. 0
			1544.19. 7½
		Profit	480.13. 4
Total	2025.12.11½	Total	2025.12.11½

One can learn a reasonable amount about the work force from the accounts. Towards the end of the year there were 6 shafts in production and assuming that each coal getting team consisted of four men, the total employment was something like:

	Coal getting	Coal transport (winding, boating, driving the horse)	Surface (banksman, coal breaking)
Buckett Engine Pit	4	4	4
Old Rise Pitt	4	1	1
New Rise Pitt	4	1	1
Goit Pitts	12	6	5

plus Thomas Wyld, in charge, John Renshaw carpenter

This gives an estimated total of 49 people. There is no evidence that any women are employed. Almost certainly the 'coal transport' column relates to boys, probably in the age range 10-14 because their day wage rates were only in the range 3d - 1/-. Day wage rates for the other jobs were in the range 1/4d - 1/8d generally, but with some day wage rates in the range 2/- to 2/6d (e.g. blasting, cleansing a sough).

The names of 72 individuals are recorded in the accounts in connection with the operation of the collieries (i.e. excluding suppliers). Some appear only once on tasks such as getting stone and they may be just casual employees rather than part of the workforce. Some appear only in one part of the year on identifiable tasks which are carried out by other people at other times. Some of the lads appeared to share jobs as sometimes 12 days of 'driving the engine horse' is attributed to one person and sometimes to two.

The names of these 72 individuals are given in Appendix 7 with their apparent main job (or named job). Since all of the entries to coal getting refer to 'Wm Wild & Co' etc, one can only assume that individuals named on casual underground jobs such as 'repairs to shaft' were normally employed on coal getting. Certainly, where there is a peak in underground maintenance there tends to be a dip in coal production. Since the estimated workforce at any one time was 49, most of the workforce must appear by name in the accounts at one time or another.

Day wage rates are given for typical jobs where appropriate. Those of 1/- per day or less are assumed to be lads, 14 or less say, and marked appropriately.

Points of interest include the strong representation of certain families, e.g. Ashmore, Bennett, Bagshaw, Kirk, Nadin, Turner, Ward and Wheeldon. None of the Wheeldons who signed the 1780 lease (see Chapter 3) - Richard, Isaac or Edmund - appear in the 5 Wheeldons listed.

Summary

Thomas Wyld's very detailed accounts give a clear insight into the workings of the Buxton collieries in 1790.

They show an enterprise with a coal production of at least 14,000 tons per year (more, if the local corf held more than 1 cwt, as is quite possible) which increased in production capacity during the year as the addition of a manpowered winder raised the number of shafts in production from 5 to 6.

Such an enterprise compares quite favourably in size with traditional late 18th century mines in other parts of the country. For instance, in one of the leading coal producing areas, the Tyne,[7] there were 21 collieries in 1767 of which 8 had a smaller production and 13 had a larger production than the Buxton collieries. At that time several collieries were beginning to increase greatly in scale with the benefits of steam power and 100,000 tons per year were being achieved in the biggest. Nevertheless, by the traditional standards, the Buxton collieries represented a medium sized enterprise. They were not very advanced technically with maximum shaft depths of about 130 yards; the most noticeable technical feature was the use of an underground canal for coal transport purposes.

The product was in considerable local demand and all of it was sold as produced to customers coming to collect it in their carts from the area bounded by Macclesfield, Chapel en le Frith, Hartington and Taddington. The enterprise also made an estimated profit of about £480 p.a. on a turnover of £2000 p.a., comparable with many of the local lead mines, although not large by the standards of the Ecton Mine which made its peak annual profit of £6,000 on a turnover of £24,000 at about this time.

The coal helped the growth of the local limeburning industry which in its turn supplied mortar for the building needs of the Industrial Revolution and lime for the Agricultural Improvements of that time. The collieries were also an important factor in local employment. About 50 men and boys worked in them at any one time. An estimate for the whole of Hartington Upper Quarter in 1788 was that it contained 130 houses, about 500 people. The 1801 Census gives 153 inhabited houses and 655 people. It is therefore likely that the population of Burbage, as a small area within Hartington Upper Quarter, was about 200 in 1790 so that the employment at the collieries probably helped to support about half the population. Working conditions would have been very arduous not only in the underground jobs but also in the surface jobs as production was maintained throughout the year and surface workers would have been exposed to some very adverse conditions of wind, rain and snow on their exposed ridge.

Whilst all this industrial activity was going on other notable local events were taking place. The fifth Duke of Devonshire, responsible for so much investment in the mines, was developing Buxton as a fashionable watering place. The Buxton Crescent was opened in 1784 and by 1790 the spa season had become well established. Incidentally, when the Crescent foundations flooded during construction, a pump from a local colliery was used to clear them.[8]

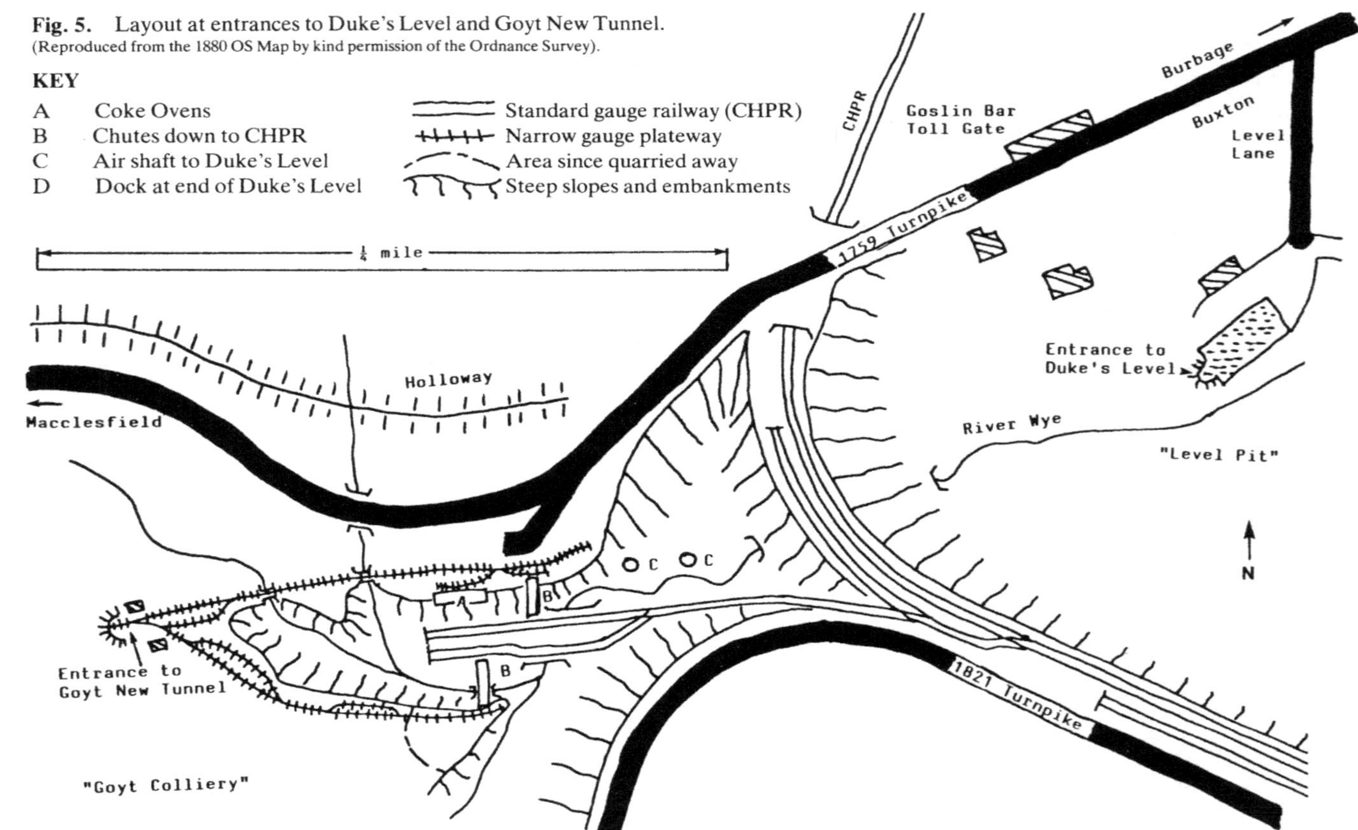

Fig. 5. Layout at entrances to Duke's Level and Goyt New Tunnel.
(Reproduced from the 1880 OS Map by kind permission of the Ordnance Survey).

CHAPTER SEVEN

LATER DEVELOPMENT OF THE COALFIELD

The first major development was the driving of the Duke's Level. This was a level tunnel used both as a drainage channel and a canal for the underground transport of coal as described in the previous chapter. It was driven straight and level at an altitude of 1,170 feet AOD for a distance of 1,360 yards, where it intersected the House Coal seam. The mouth of the Duke's Level was at the end of the present Level Lane in Burbage (Level Lane is about the only name with a mining connection in Burbage); as described earlier it terminated in a small dock and the overflow from the dock led down to the River Wye, only a few yards away. The cross section of the Duke's Level is not recorded but it was probably about four to five feet across. The cost of driving it is not recorded but as extensions to it were only paid at 1/- per yd. it was probably only about £70.

The choice of site for the mouth of the Duke's Level was mainly dictated by the lie of the land as the developers would have been seeking a low level for the entrance without an excessive length of tunnel. The chosen site avoids the River Wye and is slightly above it; to have made it at a lower level would have taken it below the level of the Wye, which was required as a natural drainage channel. The dock at the mouth of the Duke's Level is only a short distance from the 1759 Buxton-Macclesfield Turnpike and connected to it by Level Lane, as shown on Fig.5.

It seems likely that the Duke's Level was driven after the construction of the turnpike in 1759, in view of its close proximity. Indeed the opening of the turnpike would have given a powerful boost to investment in local industrial development because of the reduction in transport costs and may have prompted the driving of the Level. By 1790, the workings had extended to the Buckett Engine Pitt; therefore a date of about 1770 for the construction of the 'level' seems plausible.

The Duke's Level acted as a drainage level for all subsequent development of the House Coal seam up to its abandonment in 1919. Any coal extraction below this Level would have required a major investment in pumps or an additional lower drainage tunnel and this investment was not forthcoming.

On reaching the House Coal seam, the level was extended southwards in the seam and probably northwards as well, although this is not recorded. On the Abandonment Plan[1] it states that:-

'The water level to the dip of the workings has been surveyed to the point where it is shown on the plan. The level continues beyond this point but there are no records to show how far this level goes. It is an older working and the present owner has no record of its working. The level delivers its water into a long adit tunnel known as the 'Duke's Tunnel''.

This level is shown in Fig.3b, plus an estimated extension of it. In the above extract the 'present owner' had records going back to 1859 and the extension of the level is clearly older than that. In 1815[2] Farey recorded that the Thatch Marsh workings were 2¼ miles long. If one assumes that the level continued to the Derbyshire/Staffordshire border as shown in Fig.3b, which represents the limit of the Duke of Devonshire's coal rights, this gives a length for the level of 2⅛ miles from its mouth to its estimated termination. In view of Farey's observation, it therefore seems likely that the level had reached its limit by 1815 and had been driven in the seam about 2300 yds in 45 years, i.e. about 50 yds per year (about 30 yards of drivage were recorded in 1790, see Chapter 6).

Apart from the Duke's Level, the only other major development of the workings was the driving of the Goyt New Tunnel in about 1812. In Volume 1 of Farey's Book, compiled and written in about 1811, he refers to the proposed driving of a new tunnel at Thatch March Colliery to take a railway for underground coal transport. In Volume 3 (published in 1815) he says:

'At Thatch Marsh Colliery and at Ecton Copper Mine there are tunnels of considerable lengths, driven into the hills, in which railways are used, for bringing out the coals and ore. The Thatch Marsh tunnel has lately been driven by the Duke of Devonshire for the better supply of Buxton with coals, under the superintendance of Mr George Dickens, his colliery agent'.

The tunnel referred to here, driven in about 1810-1812, is clearly the tunnel referred to on the Abandonment Plan[3] for Goyt's Colliery as the Goyt New Tunnel. The mouth of this tunnel is also close to the Buxton-Macclesfield turnpike and connected to it by a short road; at an altitude of 1360 feet, it is at a higher level than the Duke's Level (SK07 0315.7209). This rail tunnel served as a transport road for both coal seams because it was driven horizontally through the House Coal seam and on to the Goyt seam, a distance of 3,080 yards. The route of the railway tracks can still be seen in the area between the tunnel mouth and the approach road. The visible remains show that it was a narrow gauge plateway.

Paved walkway for horses on the plateway from the Goyt tunnel.
See the wheel wear on the right hand side.

The choice of location proved to be a fortunate one because when the Cromford and High Peak railway was constructed (opened in 1831) it passed fairly close to the colliery at an altitude of 1,265 feet AOD. A short siding from the CHPR to the area below the tunnel mouth enabled coal to be loaded into trucks on the CHPR by a gravity feed arrangement, probably down wooden chutes from the level of the railway system passing into the Goyt New Tunnel. The layout of the tracks and inclines is shown in Fig.5.

The coke ovens in Fig.5 consist of a row of 4 small beehive coke ovens of a type developed in the early 18th century and superceded in the second half of the 19th century. The ovens were constructed during Mr. Hubbersty's time as Agent for the Buxton Lime Co. and the coke was used for limeburning at Grin Quarry to make better quality lime for the steel industry.[4]

Jackson[5] records that the limestone workings at Grin and the Goyt Colliery were leased to John Boothman shortly after the opening of the CHPR. The Chatsworth archives[6] record that in the early 1850s, he leased the coal mines from the Duke of Devonshire. He paid an annual rent of £700 for the Thatch Marsh and Goyt collieries and £75 for three limekilns on Grinn (i.e. Grin Low). The Memoirs of the Geological Society of Great Britain[7] include annual mineral statistics of the U.K. at that time but they are obviously incomplete and inaccurate. In 1856 they record for NW Derbyshire area:

COLLIERY	WHERE SITUATED	OWNERS
Bonsall	Buxton	J. Srigley
Brundidge	Buxton	L.and E. Hall
Furnes & Bagsworth	Buxton	J. Stott and Co.
Whaley Bridge	Whaley Bridge	Executors of Thos. Gisborne

Thatch Marsh Colliery was clearly operating at that time (at least J. Boothman was paying rent for it) and the collieries listed as being at Buxton are all north of Whaley Bridge. In 1860 there is an entry relating to Burton Colliery at Burton owned by the High Peak Rail Co. At that time Marshall[8] records that the Cromford and High Peak Railway Co. was trying to take over the limestone quarries at Grin Low and Harpur Hill (both served by the CHPR) but were encountering objections. Presumably the above entry relates to Buxton rather than Burton and the CHPR would have taken over the collieries as well as the quarries, if only for a short time.

In 1862, the entry becomes Buxton Colliery at Buxton, owned by R. Broome and Co. and this entry is repeated until 1865. In 1866 the entry becomes Axe Edge Colliery at Buxton owned by the Buxton Lime Co.

In 1872 the workings are called Axe Edge Colliery and Thatch Marsh Colliery as separate entries, but in 1874 the entry settles down to Axe Edge and Thatch Marsh Colliery, the ownership remaining with Buxton Lime Co.

By 1882, the statistics were compiled by HM Inspectors of Mines and the entry is given as Goyte Colliery and Thatch Marsh Colliery, both owned by Buxton Lime Co. and both working the Mountain Seam. H.A. Hubbersty (Agent) and William Day (Manager) are recorded for these and the Whaley, Waterloo and Horwich Tunnel collieries at Whaley

Bridge. H.A. Hubbersty, as Agent for the Buxton Lime Co, initiated the amalgamation of 13 local lime firms into the Buxton Lime Firms Co. Ltd.[9]

At the turn of the century William Harvey was manager of Buxton Lime Firm Co. with collieries at Thatch Marsh and Whaley Bridge.

Local names for the shafts in the Thatch Marsh complex were 'Back o' th' Moss', whose coal was conveyed to the Cromford and High Peak Railway by boat, and the 'Sluther', 'Crashaway' or 'Moss' pit which had an engine. However, water was drained by the 'level' to Burbage.

Both Farey[10] and Glover[11] refer to 'brasses' (iron pyrites) in the coal seam at Thatch Marsh. These were extracted and sold for use as 'Copperas', 'Green Vitriol' or 'Sulphate of Iron'.

In 1850 a law was passed requiring proper plans to be kept of mine workings and from 1859 onwards there are complete records of the workings at Goyte Colliery and Axe Edge and Thatch Marsh Colliery.

These plans form the basis of the Abandonment Plans, required by law to be lodged with HM Inspectors of Mines when coal workings are closed down. These plans are now held by the East Midlands Records Office of the NCB.

Plan 3072 records that the Goyte Mine, Burbage, was abandoned on 31/12/1893, the reason given being

'Area above water level worked out'

The owners were given as the Buxton Lime Firm Co. and the seams worked as the Goyte's Seam and the Mountain Mine Seam.

Plan 6915 records that Thatch Marsh Colliery was abandoned on 7/6/1919, the reason given being

'coal worked out down to the water level'

The owners are recorded as the Buxton Lime Firm Co. Ltd, Royal Exchange, Buxton.

These plans record the areas worked during successive 12 month periods from 1859 to 1921 and are therefore very detailed. They also record the edges of 'old workings', i.e. areas worked prior to 1859. This information has been used to prepare Fig.3b which highlights estimated regions for the pre-1859 workings and subsequent workings in labelled blocks.

For the House Coal seam the picture is fairly straightforward. As stated earlier, the water level was extended probably down to the county line and a limited amount of extraction would have taken place during this work; the bulk of the early extraction though was from the fault near Berry Clough for about 2000 yards southwards. The post 1859 extraction was in the vicinity of a 112 yard deep shaft on Axe Edge shown as 'working pit' on Fig.3b (SK07 0270.7055), but then switched to the vicinity of a 1 in 6 drift sunk from the Cistern's Clough area. From 1871 onwards there was a fairly systematic extraction of the reserves from here southwards to the county line. By the beginning of the 20th century, this extraction was really a mopping up operation of the available reserves. Some attempts were made in 1916-18 to develop new areas but these ran into faults and were largely unsuccessful.

The drift was apparently sunk about 1870 although it is not shown on the 1880 25 inch O.S. map of the area. It is shown on the 1898 edition with a rail system running down an incline to small buildings near the large loop at Cistern's Clough in the main Buxton-Leek Road (this loop has now been eliminated from the road by building a large embankment but it remains as a car parking area). The Ordnance Survey refer to these workings as Burbage Colliery. Locally it was known as 'Top Pit'.

Remains of the 'Engine House'. (See fig. 3b).
(Reproduced by courtesy of Derbyshire Museum Service)

The rails extended down the drift into the workings and there were small buildings where the drift system and the rails on the incline met. There was a steam powered winding engine there serving both systems; its foundations and a small pond are still visible. The bottom of the incline terminates in an area where carts would have been loaded, a short distance from the main road. An endless rope brought small wagons of coal from the pit bottom and conveyed them down the incline to be tipped on to the 'iron'. An office was situated at Cistern's Clough.[12]

By the time of the survey for the 1922 edition of the map only a short length of rails remained on the level ground near the top of the incline and into the drift. Possibly the rail system had fallen into disuse before the colliery was abandoned. With the advent of motor powered lorries it would have been almost as easy to drive to the drift entrance for loading as to drive to the bottom of the incline. A further inducement to abandon the incline might have been winter floods. The incline ran more or less down the line of Cistern's Clough but crossed it three times on bridges which, judging by their remains, were not very substantial. It may have been a considerable chore keeping these bridges in repair and the track in operation during the winter.

Post-1859 operations from the Goyt Tunnel included a small amount of extraction in the House Coal seam, but the rest of the extraction in this period and from this tunnel was in the Goyt Seam. Here the pre-1859 workings had extracted most of the central area of the seam, which was more or less level, and the post-1859 workings were left with the sloping areas leading up towards the outcrop. The area near the end of the original Goyt Tunnel was extracted mainly from 1862-1880; a further major block of reserves at the far side of the seam was extracted via an extension of the Goyt Tunnel, in the period 1870-1889. It was not recorded on the Abandonment Plan whether the old workings extended into the north west corner of the Goyt seam, i.e. north of the Turnpike and west of the River Goyt. That apart, the entire seam appears to have been thoroughly exploited.

There are no drainage levels from the Goyt seam shown on the Abandonment Plans. However, Cope[13] refers to a cross measure drift from the Goyt Valley into the seam and there is certainly the mouth of an artificial tunnel about 6 feet wide and four feet deep issuing into the Goyt Valley at SK07 0178.7207, at an altitude of 1285 feet AOD. This in the right place to serve as a drainage level for large areas of the Goyt seam and certainly one would expect such a level to be necessary in exploiting a shallow seam in a marshy area. Indeed, pencil notes on an original mine plan held by the Peak Park Planning Board record the water flows in different parts of the Goyt seam workings. A total of 125,000 gallons per day is recorded.

The Abandonent Plans for both seams record a number of Air Pits as well as other shafts. These plans do not generally show surface installations and there are no references to fan installations. The O.S. maps show the details of buildings etc. but do not name any fan houses. It seems likely that the entire operation, extended over considerable distances, relied on natural ventilation throughout its lifetime, provided either by the appreciable wind effects (particularly with shaft outlets at different altitudes and on different directions of slope) and the thermal effects that derive from the difference between the air temperature and the temperature of the strata.

Sough draining the 'Goyt' seam. Situated in the Goyt valley near Derbyshire Bridge.

An additional plan at the East Midlands Mining Records Office refers to 'Plans of Goyt Moss Colliery and Level Colliery' belonging to the Buxton Stone Co. It also refers to the seams as the Level coal seam and Goyts Big Mine seam. The tunnels are referred to as the 'Duke's Old Level Tunnel' and the 'New Tunnel'. The entrance of the latter is referred to as the 'Day Eye' - a phrase highly evocaive of a long working shift in very poor lighting conditions ending as one walks along the tunnel towards a tiny pinprick of light that slowly grows until one emerges into daylight - the Day Eye.

Production statistics are also recorded on the plans in terms of areas of extraction in Acres Roods and Perches. One acre of coal in a 4 foot thick seam, as in these mines, weighs about 6400 tons and assuming an extraction rate of 80% gives one acre as yielding 5000 tons of coal.

The recorded areas of coal extraction for the two seams are given in Appendix 8 on a year by year basis together with the estimated equivalent tonnage based on 1 Acre — 5000 tons.

It can be seen that at the beginning of the period, from 1/1/59-30/6/62 (2½ years) the average rate of production in the Goyt seam was only 9100 tons/yr, rising to 14200 tons/yr in year ending 30/6/63. Thereafter, production in the Goyt seam remained in the range 19000-27000 tons/yr until 1883. From 1883-1888 it was in the range 12900-22500 tons/yr but thereafter it dropped rapidly. 1893 showed some revival, to 22700 tons/yr, but this was a last fling before the official abandonment on 31/12/93, mainly in a pillar of coal left at the inner end of the Goyt Tunnel. Apparently, some coal production continued after the official Abandonment because small areas were worked each year from 1894-1898.

The House Coal seam had an altogether more modest output in this period. In the first 2½ year period, it averaged 7100 tons/yr but it only once exceeded this figure, with 7400 tons/yr in 1873. After 1874, production was consistently low, around 2-4000 tons/yr dropping to about 1000 tons/yr by 1898. Production areas are only quoted until 1902 but the areas worked are still shown on the plan until abandonment of Axe Edge and Thatch Marsh Colliery in 1919. The period from 1903-1919 produced an estimated 20,000 tons of coal.

The total estimated production from 1/1/59 was 653,800 tons for the Goyt seam and 167,000 tons for the House Coal seam.

Estimated production prior to 1859, based on the areas given in Fig.3 for extraction via the underground workings are 200,000 tons and 300,000 tons for the Goyt and House Coal seams respectively. In Chapter 5, the number of small pits recorded is 130 for the House Coal seam and 108 for the Goyt Coal seam; each of these pits could have produced about 2000 tons of coal.

An overall production estimate for the coal mines would therefore be:

	Goyt	House Coal
Extracted via small shafts	216,000	260,000
Extracted via underground workings (pre 1859)	200,000	300,000
Extracted via underground workings (post 1859)	653,800	167,000
TOTAL	1,069,800	727,000

In Chapter 8 it is noted that 35 coal miners were recorded in Burbage in 1881. The estimated production in that year was 23,300 tons giving an output per man of about 700 tons/yr. The overall estimate for the whole of Derbyshire for that year was 380 tons/yr per underground worker, so that productivity at Burbage was well above the county average even allowing for possible appreciable errors in the estimates.

The District Inspector for the Midlands, Mr.A. Stokes, described the mining industry in that area as having changed little in techniques since Anglo-Saxon times. Although an appreciable exaggeration, it does make the point that modern methods based on steam power, explosives, and electricity came slowly to the many low capital investment mines of the country. The Buxton coal mines changed little in technology from 1790 to their eventual closure and, apart from the use of a stationary steam engine for haulage at Cistern's Clough and horses for coal winding and coal transport elsewhere, they relied mainly on man power throughout. Their peak of technical innovation was probably the use of canal transport in the late 18th century. Thereafter, the U.K. coal industry moved on apace and left the Buxton collieries, and many others, with an essentially 18th century technology. Nevertheless, the mines produced about 1.8 million tons of coal at competitive prices (worth about £60 million at current pithead prices) and this activity had an important effect on the development of the economy of the region.

With improved rail access, cheaper, better quality coal became available. The quality of the local coal did not justify large scale investment to permit mining below the 'Duke's Level' because of the water problem.

In times of national emergency the mines were reopened for brief periods. In the General Strike, 1926, coal was supplied from the Axe Edge mines to the 'Magpie' lead mine.[14]

In the central area of the coalfield, at Ludworth, a one-man working mine still operates.

CHAPTER EIGHT

THE MINERS

Occasional glimpses of the men who worked the coal mines survive in various documents (no record of any women working in these mines has yet been seen).

In 1780, Robert Longdon, Isaac Wheeldon, Richard Wheeldon and Edmund Wheeldon feature in the lease with the Duke of Devonshire. Farey[1] describes Robert Longdon as the agent of the Duke of Devonshire in 1783, together with George Brassington.

In 1790 a clear idea of the workforce is given by Thomas Wyld's accounts, as described in Chapter 6. These accounts identify 72 people working in the mines or connected with the mining operations.

Farey records that in 1815, George Dickens was the Duke of Devonshire's Colliery Agent and the Chatsworth[2] archives record that in the early 1850s J. Boothman leased the mines from the Duke of Devonshire. By 1882, official collection of statistics was well under way and H.M. Inspectors of Mines records show that H.A. Hubbersty was the agent and William Day the manager for all the Buxton Lime Firm collieries at that time, recorded as Goyte, Thatch Marsh, Whaley, Waterloo and Horwich Tunnel, the last three being near Whaley Bridge. Local Directories of the time simply record Mr. Hubbersty as a 'Gentleman': perhaps they did not wish to advertise the presence of coal mines close to a burgeoning spa.

H.M. Inspector of Mines statistics also record that 5 miners were killed in accidents at Thatch Marsh Colliery between 1865 and 1889; this figure may be misleading because in 1889 a fatality is recorded for Thatch Marsh Colliery in the statistical tables which is elsewhere recorded as the death of Charles Foulstone in a fall of ground at Whaley Colliery (also a Buxton Lime Firms Colliery). It seems likely that the 5 recorded fatalities relate to the 5 Buxton Lime Firms Collieries in that period.

Further information is available from Census data; census information on individuals is available from 1841-1881 and information on those with an identifiable connection with the coal mines is summarised in appendices 9-11 for the Censuses of 1841, 1851, and 1881.

Thomas Wyld's accounts (Chapter 6) identified the Ashmore (5), Bennett (4), Bagshaw (3), Kirk (4), Nadin (3), Turner (5), Ward (3) and Wheeldon (5) families as making a substantial contribution to the workforce in 1790, the figures in brackets indicating the number of individuals identified in each family.

The 1841 census returns show a striking similarity with the 1790 records. Out of 28 people identified as working in the mines, there are 6 Ashmores, 6 Bagshaws, 6 Bennetts and 3 Wards. The 1790 accounts also record John Street and there are 2 Streets in the 1841 records. The other names differ between 1790 and 1841. Thus the Ashmore, Bagshaw, Bennett, Street and Ward families contributed 16 miners in 1790 and 23 (out of 28) in 1841. The forenames also repeat themselves. Between the 1790 and 1841 records, there are the following names in common: George Ashmore, Thomas Ashmore, Charles Bagshaw, John Bennett, William Bennett, John Street, Edward Ward, John Ward and Samuel Ward. The ages in the 1841 census records show that none of the miners identified in 1841 was of working age in 1790 so the repeated names refer to different individuals.

In 1841, the Ashmores lived at Gate Farm and at Dog Holes (both of which are near where the Macclesfield Old Road crosses the River Wye) and at Germany (now known as Holmfield); the Bagshaws lived at Shay Lodge, the Bennets at Handcroft (now known as Anncroft), the Streets at Level House (adjacent to the entrance to the Duke's Level) and the Wards at Goyt Moss (near where the 1759 turnpike crosses the River Goyt) and at Gutter House.

In 1841 records show one 12 year old (Charles Bagshaw), three 15 year olds (Charles and James Bagshaw, Edward Street) and one 16 year old (Thomas Bagshaw) as working in the mines.

John Street (together with his son Edward) later became a 'banksman' at the 'Level Pit'.

In 1851 the picture was similar to that for 1841. The identifiable labour force had risen from 28 to 31 with the Ashmore (6), Bennett (5), Bagshaw (1), Street (3) and Ward (4) families contributing 19 of the 31. Significant newcomers were the Norton (4) and Wharmby (4) families. The 1851 census for the first time recorded information on place of birth: of the 31 individuals listed, 29 were born in Burbage or Buxton. David Ward, resident at Goit's Moss, was born in Macclesfield Forest. Part of Goit's Moss was then in Cheshire but the Census Returns for the Parish of Taxal, which covered this area, only list colliers at the Fernilee and Whaley Bridge end of the Parish. (Some members of the Ward family lived across the county line in the Parish of Taxal portion of Goit's Moss but they were listed as farmers). In 1851 therefore, the mines were worked by a labour force that was essentially local in origin.

Between 1841 and 1851 the Ashmore family had split up to some extent. John Ashmore, with his son William, had moved from Dog Hole to Wye Head. James Ashmore had moved to Axe Edge. Charles Bagshaw had moved from Shay Lodge to Handcroft where the Bennetts had remained.

The number of young people working in the mines had increased. The 1851 census returns show a 12 year old (Samuel Norton), a 13 year old (Samuel Wharmby), three 14 year olds (William Ashmore, Joseph Bennett, William Norton), a 15 year old (John Wharmby) and a 16 year old (Josh Norton).

By 1881, the picture had changed somewhat. The workforce had increased to 35 with one recorded unemployed miner (William Bennett, aged 71). The old village families were still represented by the Ashmore (4), Bagshaw (2), Bennett (4) and Street (1) families but this time totalling only 11 out of 36. Ninety one years after the 1790 records, there were still some of the same names repeated - Thomas Ashmore, Charles Bagshaw, William Bennett.

There were fewer young people in the mines in 1881, with only one 15 year old (John Goodwin) and two 16 year olds (George Brunt and ——— Bowyer).

A notable feature is the increase in the number of people born elsewhere. There are John Pine and John Price, both from Devonshire and George Packwood from Worcestershire; from closer at hand there are several people who were born in nearby areas that also had small coal mines - Quarnford, Flash, Whaley Bridge, Macclesfield, Pott Shrigley, Wildboarclough. By 1881 the law required a manager with a certificate of competency to be appointed and William Day is recorded as the manager. He was born in Heanor, as also were John Hay and Francis Norton.

The workforce was becoming more spread out in 1881. Harper Hill (now Harpur Hill), Grin Row and Boothman's Cottages appear as addresses.

All of these records, taken with the general picture of production in the mines, suggest that 1790 was a time of fairly rapid expansion with people being drawn into the Burbage area to meet the labour needs of the mines; by 1841 the workforce was down to 28 and was completely dominated by local families who had been there in 1790 if not before. A local shortage of labour may have been significant from the 1850s onwards as by 1881 members of the old village families, although still represented, were in a minority and employees had been drawn in to the village, some from considerable distances.

The Goyt Colliery closed in 1893 and the production records show that Axe Edge and Thatch Marsh Colliery was a much smaller affair.

Employment probably reduced into the range 10-15 people and, because the colliery was being worked from the Cistern's Clough end, the employees might have come from the Brandside/Axe Edge area thereafter.

Burbage Colliery closed in 1919. Personnel in the final years were 'getters' — Moses Bagshaw, Ned Goodwin and Joseph Shufflebotham, 'banksman' Joe Findlow and 'engineman' Will Heathcote.[3]

In 1790, coal mining would have been the main source of employment in Burbage, helping to support about half the population. Burbage was never a mining village in the conventional sense of long terraces of houses built specifically to accommodate miners. Rather, the miners lived in scattered cottages and small farms many of which were there before mining started; the mining families probably ran smallholdings as well in many cases. During the 19th century, the population of Burbage grew while the mining industry remained relatively static in size. By the time of the Goyt Colliery closure in 1893, the village had developed sufficiently to be fairly unharmed by the loss of jobs. As today, people who had moved long distances to get work, from Devonshire for example, were probably most affected, having been in the village long enough to put down roots but still being regarded as newcomers.

CHAPTER NINE

INCIDENTAL MINING CONNECTIONS

The geology of the Buxton area and the local coal mines give rise to some perhaps unexpected mining connections:

Writing in 1886, Dr Robertson[1] (a long time medical superintendent of the Devonshire Hospital) describes how experiments were in progress to treat the local sewage with a mixture of powdered lime and the water drained from the coal mines at Burbage; on their passage through the ground and into the workings the water dissolves iron salts and carries others in suspension, giving it a rust coloration. The experiments were successful and a pipeline was constructed to take water from the Duke's Level to a treatment plant in Ashwood Dale (a distance of about 2½ miles).

The water from the Duke's Level has also been used in times of drought by pumping it into Burbage Reservoir.

Slightly north of Buxton John Shallcross, of Shallcross, as owner of the 'High Peak Coal Mines' issued 'trader tokens'.[2] One of these tokens is in the possession of Buxton Museum and Art Gallery.

Although working of the coal seams officially finished in 1921, local tradition has it that coal was extracted unofficially during the General Strike of 1926 and again during the severe winter and prolonged fuel shortage of 1947.

A further local tradition is that after 1945, a substantial quantity of army surplus electrical equipment was disposed of by dumping it down a local coal mine shaft; an enterprising local business man rented the ground surrounding the shaft and recovered the equipment (quite legally) to start a successful 'army surplus' shop.

The Grinlow Quarry ceased working during the 1950's and at a time of considerable national overproduction of coal, following a major growth in oil imports, the entire excavated area was filled with part of the national stockpile of coal (about 1 million tons of it). The coal was clearly visible above the rim of the quarry and traces of it can still be seen around the quarry edge (to confuse the amateur student of geology). The coal was removed during the 1960's and the quarry operated again briefly before being finally abandoned. A major reclamation scheme in the 1980's recontoured the tips and created the present layout.

During the 1880's the death toll in mining accidents exceeded 1000 per year in the United Kingdom, with over 200 deaths a year on average from mine explosions. The concern created by this accident rate focused national concern on mining safety and a major mine explosion in France in 1903 in which 1100 miners were killed gave considerable impetus to demands for research into the causes and control of mine explosions. Systematic research on mine explosions first took place at Altofts Colliery in Yorkshire and then at Eskmeals in Cumberland. In the early 1920's a more suitable site was being sought, more central to the main mining areas of the country where an explosion gallery 1200 feet long could be erected on a straight level piece of ground in a fairly remote area where the noise of the test explosions would not cause undue disturbance. The search revealed an ideal site at Harpur Hill, to the south of Buxton in a sparsely populated area with the (for the time) massive quarries of Harpur Hill and Grin on either side, already generating noise from blasting; a diversion of the Cromford and High Peak Railway in 1875 from its original 1831 route had left a disused stretch of track with exactly the features for installing the explosion gallery. So by a rather long chain of circumstances, the Safety in Mines Research Establishment came to Buxton where it still operates today as a laboratory of the Health and Safety Executive. Larger cross section mine roadways and more modern mining equipment can be seen there on the surface than ever existed in the real coal mines beneath Axe Edge.

As part of its work, the laboratory issues certificates confirming that electrical equipment meets certain rigorous standards essential to its safe use in mines. For many years the description 'Buxton Certified' appeared on advertisements and trade literature of mining electrical equipment throughout the world as a guarantee of its safe design; the 'Buxton Certified' description is far better known in mining circles than ever the real Buxton coal mines would have been!

Finally a few words about two other types of mining, which have largely gone unrecorded, in the Burbage and Grin areas of Buxton. Although not directly concerned with coalmining, they obviously played a small role in the local economy and it is hard to imagine in such a small community that there was no exchange of mining techniques or even interchange of employment by the miners themselves.

In the Burbage, Ladmanlow, Grin and Stanley Moor areas, a few lead mines were worked in a very small way. The geological map shows five very short veins in the 'Bee Low' limestone and also within these veins were calcite and barytes. One of these veins was called the 'Sough Gate vein' which probably had some connection with a sough which drained out close to Ladmanlow toll bar. Three of the mines were known as 'Grin End', 'Solomons Temple' and 'Watricle'.[3] The census of 1851 records one Mr. Doxey a 'lead miner' of Harpur Hill.

The barytes, or 'caulk' as it was commonly known, was also extracted from the veins. It was commonly used at one time in the paint industry and as a filler in the paper industry. Its use has declined with the advent of synthetic paints, but barytes mills at Goytsclough (SK07 0120.7317) and Barmoor Clough (SK07 0722.7990) attest to its former importance.

Little is known about these mines but circa 1843 a 'caulk' mine collapsed entombing John Bagshawe and Benjamin Bonsall from 11.00 hrs Monday to 08.00 hrs Saturday. The Buxton Lime Company owned the mines from 1868 to 1873 when in the latter year they produced 713 tons of ore.[4]

Abandoned mineshafts present danger even when sealed. Dr Hanna of Buxton discovered this in 1916 when a sealed 'caulk' mine shaft along Stables Road, Ladmanlow, collapsed underneath his horse. Dr Hanna escaped unharmed but the horse had to be destroyed.

Not connected with above, but for the sake of record, two other lead mines close to Buxton were at Tom Thorn (SK07.0710.7570) and Great Rocks Farm (SK17 1035.7422).

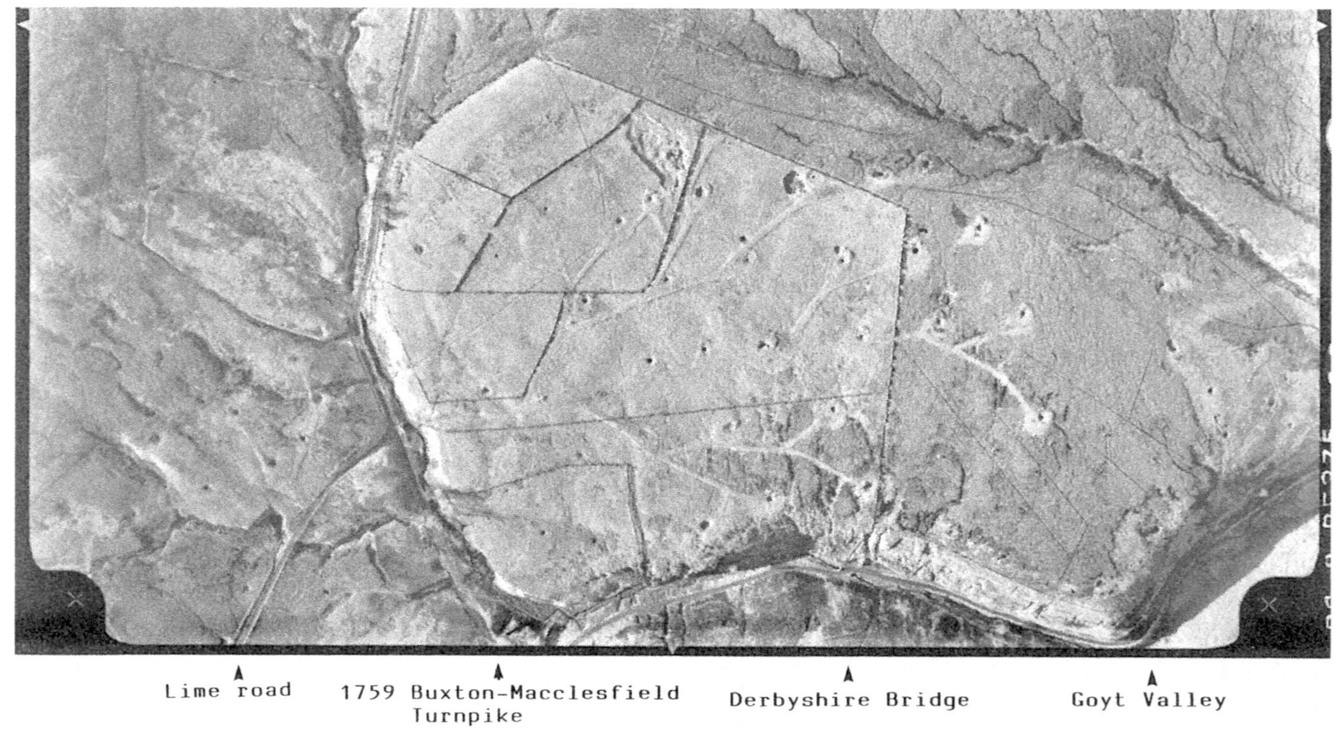

Fig. 6. Aerial photograph of part of Goyt coalfield (see Fig. 3A) showing early pits (right hand side) and later pits with access roads. (Cambridge University Collection: copyright reserved).

CHAPTER TEN

THE COAL FIELD TODAY

Sixty years after the last official mining of coal, surprisingly little remains as visible evidence of the whole enterprise. The surface installations have nearly all been removed, the shafts and drifts sealed and the moorland is reclaiming its own.

Nevertheless, traces are still there and the following description encompasses the remaining visible evidence for the area described in the earlier chapters.

NOTE: public rights of way are identified on the Ordnance Survey 1:25000 map. Although all major shafts have been sealed or fenced, there are many minor shafts in the area and these should be avoided on safety grounds.

Remains of a coke oven at Goyt colliery.

The greatest concentration of visible remains is near to the Macclesfield Old Road. Standing near Rock Bay Garage on the Buxton-Leek road and looking across the valley, a small coal tip at the end of Level Lane can be clearly seen, to the right of the reservoir walls. In Level Lane itself, the dock at the end of the Duke's Level has been filled in and now forms the garden of a private house on the site of the colliery buildings; the rectangular shape of this plot of land still indicates the area of the dock. The public footpath leading down to the River Wye and past the boundary of the garden crosses the tip and then crosses the discharge point where the Duke's Level still drains into the River Wye; this discharge is often a reddish colour from percolation through iron bearing strata deep in the hillside.

The footpath then leads up on to the embankment of the now disused Cromford and High Peak Railway. The route of the siding to Goyt Colliery is visible in the valley descending approximately to stream level. The spoil heaps from an air shaft of the Duke's Level are also visible beyond the CHPR embankment.

Slightly further up the Macclesfield Old Road a road runs off to the left which leads directly to the mouth of the Goyt Tunnel (also blanked off). Between this road and the slope down to the river, there are the remains of coke ovens and foundations near the top of the incline. Supports for the incline down to the CHPR siding are still visible in the coal covered slope leading down to the river; these suggest a chute arrangement from the top of the slope to the siding. Opposite this spot a second siding from the tunnel served another incline which has been removed by the later quarry.

Passing further along the road, some old paved stone that formed part of the track of the plateway from the Goyt Tunnel can be seen. Bridges or culverts across the cloughs leading down to the River Wye have been washed away but the route can still be followed on to the level ground in front of the Goyt Tunnel entrance. This entrance appears to have been blocked by bringing down some of the hillside across it. An ill defined footpath leads from this spot to the Old Engine Pit and on to Boothman's Cottages (site of) but it is easier to return to the Old Turnpike. Above the entrance to Goyt tunnel and further up the hill is the sealed top of one of the tunnel's airshafts. (SK07 0295.7210)

Further up the old road a footpath or track to the left runs parallel to the outcrop and serves as a transport road for the workings. It leads past a number of capped shafts, with the uneven ground created by the large number of small shafts visible in the direction of Buxton. This track terminated at the Buckett Engine Pit (see Fig.3b). Before reaching this point on the left a shaft, once sealed and now collapsed, is the remains of the 'Old Engine' pit (SK07 0280.7172).

In the middle section of the workings on Axe Edge, between the present Buxton-Macclesfield Road and the minor road to the south, the picture is similar. The tracks run parallel to the outcrop with the remains of many old pits still visible. The old shaft at SK07 0270.7055, with its accompanying tip, represented one of the major access points to the Thatch Marsh Colliery. The lined shaft at SK07 0270.7093 was an airshaft for the 'Duke's Level'.

Pool formed in a collapsed shaft near Derbyshire Bridge.

At Cistern's Clough the disused loop of the Buxton-Leek Road forms a convenient car park and from here one can see the bottom of the incline that ran up to the 1 in 6 drift into the House Coal seam. The public footpath from this car park also formed part of the access to the coal field and following this path up to the minor road (the Coal Road of the 1804 Hartington Enclosure Award) one can see the route of the incline on the left. The entrance to the drift is sealed but a small tip and the foundations of the winding house remain.

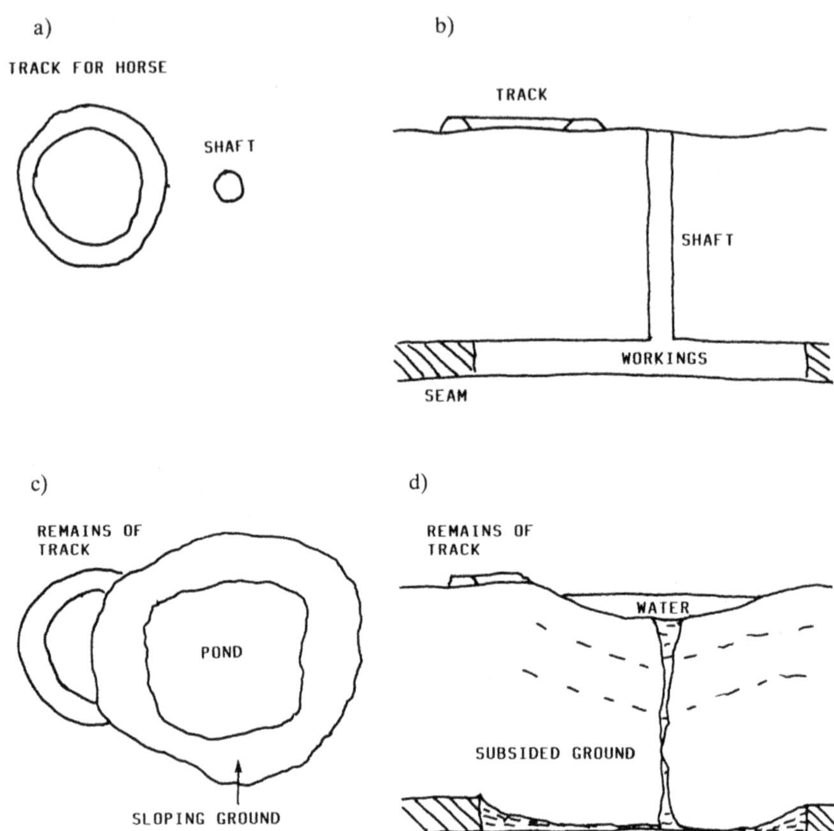

Fig. 7. Typical layout of whim gin.

a) plan view, operational condition

b) elevation, operational condition

c) plan view, present day condition

d) elevation, present day condition

Further on, a footpath to the left leads down to the Staffordshire border and the southern extent of the workings. At that point the workings are continuous back to the end of the Duke's Level and one is 2¼ miles from the Level entrance in Level Lane.

The remains on the Goyt seam are again fairly scanty but several lengths of road leading to the numerous round small pits can be seen. (See Fig.6). On the 'high' side of these pits the 'gin circles' can be seen, often partially collapsed into the shaft. (See Fig.7). The shaft and associated spoil heap at SK07 0245.7185 is the second airshaft along the 'Goyt Tunnel'.

The major remains are really the footpaths and tracks themselves; all the network of paths in this region relate to the coal workings in some way. In addition the holloways described in Chapter 3 are still clearly visible. At SK0178.7207 the entrance to a sough which drained the Goyt mines can clearly be seen.

One last point of interest relates to the miner's houses.

The footings of old cottages (Moss House, Moss Hall and Goyt's Moss Farm) near where the old turnpike crosses the River Goyt are still visible but the supreme location for anyone seeking a room with a view was Boothman's Cottages (see Fig.3b). Now demolished, these cottages were located at about the same altitude as the 'Cat and Fiddle' at 1,690 feet AOD. The views across to the east from this spot are superb but living conditions there in winter must have been severe in the extreme.

Boothman's Cottages. Erected for colliery use probably in the 1850s.
(Reproduced by courtesy of Derbyshire Museum Service)

Final Note

In conclusion perhaps we could enter a plea for recognition of the scale of the Buxton coal mines and their contribution to the local economy.

Although relatively little remains in the way of structures there are many signs of past activity. At the entrance to the 'Goyt New Tunnel', itself a circa 1812 tunnel with plateway, one is very close to an 18th century canal tunnel, the Cromford and High Peak Railway, 1759 and 1821 turnpikes, a toll bar and a number of early coke ovens. The latter are deteriorating and suffering from vandalism.

The site is important for its concentration of industrial archaeology and is worthy of preservation and recognition.

Acknowledgements

In compiling this work thanks are due to:-
Mining Records Office, N.C.B., Eastwood
Trustees of the Chatsworth Settlement (Archives)
John Rylands University Library, Manchester
Derbyshire County Record Office
Derbyshire Museum Service
Peak Park Planning Board
University of Cambridge Committee for Aerial Photography
Buxton Library
Dr. M. Bishop (Buxton Museum and Art Gallery)
Mr. A. Poulson
Mrs. A. Phillips
Mr.& Mrs. G. Barton
Mrs. J. Robinson

Appendix 1: Local Coal Pits

The names and locations of those in the immediate area around Buxton (roughly Whaley Bridge to Macclesfield to Flash) as given by Farey are:

Adlington	$2\frac{1}{2}$m SSW of Disley
Bakestone Dale	NE of Pott Shrigley
Berristow	$\frac{1}{2}$m SE of Pott Shrigley
Birchen Booth	$1\frac{1}{4}$m W of Flash
Blackclough	$\frac{1}{2}$m NW of Flash
Blakelow	SE of Macclesfield
Blue Hills	N of Bramcote, 3m SSW of Flash
Bollington	NNE of Macclesfield
Bugsworth	NW of Chapel en le Frith
Castedge	SE of Jenkins Chapel, 3m S of Taxhall
Chest (near Flash)	3m E of Wincle Chapel
Cliff Bank	E of Macclesfield
Combes Moss	2m N of Buxton
Cowpasture (near Beard)	$2\frac{3}{4}$m NW of Chapel en le Frith
Dane Head	2 NW of Flash
Dane Thorn	$\frac{1}{2}$m W of Dane Head
Diamond Hill	$1\frac{1}{4}$m NW of Flash
Diglee (or Whaley Bridge)	2m SE of Disley
Eastborough Lane	NE of Macclesfield
Ferneylee	$3\frac{1}{4}$m WSW of Chapel en le Frith
Furnace Clough	N of Whaley Bridge, $1\frac{3}{4}$m E of Disley
Gap Sitch	$\frac{1}{2}$m WNW of Taxhall
Gold Sitch	$1\frac{1}{2}$m SW of Flash
Goyte Moss	E and N of Moss Houses, $2\frac{1}{4}$m W of Buxton
Harrop	SWof Brink, $1\frac{1}{4}$m ESE of Pott Shrigley
Hayclough	near Midgley Gate, 2m E of Wincle Chapel
Hazle Barrow	$2\frac{1}{4}$m SSW of Flash
Hurdesfield	N of Macclesfield
Kerredge	$\frac{1}{2}$m W of Rainow Chapel
Latch	NW Cutthorn Farm, $2\frac{3}{4}$m NE of Wincle
Macclesfield Common	1m ESE of Macclesfield
Mouse Trap	WSW Spire Bent Farm, $3\frac{3}{4}$m NE of Wincle
New Post	$\frac{1}{3}$m N of Rainow Chapel
Notbury	1m NW of Flash
Penny Hole	$1\frac{1}{4}$m NNW of Flash
Pott Hole	$\frac{1}{4}$m NE of Pott Shrigley
Quarnford	$2\frac{1}{4}$m E of Wincle
Rainow Low	$\frac{1}{4}$m N of Rainow
Riley Clough	Macclesfield Common, 1m E of town
Robins Clough	NE of Midgley Gate, $2\frac{1}{2}$m E of Wincle
Shall cross	E of Taxhall, $2\frac{3}{4}$m WSW of Chapel en le Frith
Shrigley Fold	NE of Macclesfield
Spons	$1\frac{1}{2}$m E of Pott Shrigley
Spons Moor	$1\frac{1}{4}$m NE of Pott Shrigley
Styperson	NW of Pott Shrigley
Swanco	2m NE of Macclesfield
Thatch Marsh	in Hartington, 2m SW of Buxton
Throstle's Nest	2m E of Macclesfield
Whatshaw	$2\frac{1}{2}$m SSW of Flash

Appendix 2: An example of T. Wyld's accounts

Thatch Marsh and Goite Collierys a fortnights reconing ending the 12th day of March 1790

Ready money	Score	Corves	Price	£. s.d
Buckett Engine	144	0	at 4/2	30. 0.0
Goite	7	6	at 2/-	14.7½
				30.14.7½

		Days	Score	Corves	Price	£ s d	£ s d
Buckett Engine							
By Wm Wild & Co	getting coal		128 —	5	at 1/4	8 11 0	
By Wm Bennett	Banking do	12			at 1/9	1 1 0	
do	coals burned on bank			10	at /2½	2 1	
do	a man assisting on do	3½			at 1/6	5 3	
By Saml Bennett	Banking	12			do	18 0	
By Jno Hibbert	Brake'g coal	12			at 1/4	16 0	
do	waite'g over the men at night					2 6	
By Thos Kirk	Fill'g and Hook'g of coals	12			at 1/10	1 2 0	
do	Repair'g the Navygation	½			at 1/6	9	
By Wm Norton	boating coals	13			at 1/-	13 0	
By Wm Kirk	do do	13			at /8	8 8	
By Jacob Wheeldon	Lad'g water	12			at /8	8 0	
By Josh Renshaw	assisting to boat coals	6			do	4 0	
By Saml Trafford & Co	Cleans'g of Lodge and Land'g					8 2	
By Jno Renshaw	Carpenter	12			at 1/6	18 0	
do	repairing the boats at night					3 0	
By Wm Longson	mend'g of roads	4½			at 1/2	5 3	16 6 8
The New Rise Pitt							
By James Belford & Co	sinking 20 yards				at 4/6	4 10 0	
do	getting stone out of the pitt top					10 0	
do	making three Flakes					4 6	
do	geer'g the Pitt					2 6	
do	setting 15 pairs of timber				at 1/-	15 0	
do	drawing timber	2			at 1/6	3 0	6 5 0
Goit No. 5							
By Robt Bagshaw & Co	getting coals		170 —	2	at /10	7 1 9	
do	rais'g shael					2 6	
Goit No. 3							
By Jas Hudson & Co	getting coals		104 —	11	at do	4 7 1½	
do	Drive'g of endway 10 yards				at /6	5 0	
Goit No. 4							
By Wm Vawdrey & Co	getting coals		143 —	6	at /10	5 19 5	
do	send'g shael	6			at 1/2	7 0	
By Jno Ward & Co	Bank'g coal	48			at 1/6	3 12 0	
By Geo Turner & 3 others	Drive'g the Engine Horse	36			at /6	18 0	
By expenses of going to Macclesfield to buy Powder for the works						2 6	22 15 3½
							45 6 11½

"Thomas Wyld's accounts of a fortnight's reckoning of the Buxton Collieries" 1790

Appendix 3: OUTPUT FROM BUXTON COLLIERIES DURING 1790 (score/corves)

Fortnight ending	Buckett Engine Pitt	Old Rise Pitt	New Rise Pitt	Goit No 1	Goit No 2	Goit No 3	Goit No 4	Goit No 5	Goit No 6	Goit No 7	Goit No 8
15 Jan	116—11	86—8		145—17		125—13	51—11				
29 Jan	129—14	102—12		116—7		88—18	133—3				
12 Feb	128—15			166—5		151—8	138—11				
26 Feb	132—6					111—3	115—10	126—14			
12 Mar	128—5					104—11	143—6	170—2			
26 Mar	97—10					51—16	59—11	157—9			
9 Apr	115—7	71—3			113—15			125—9			
23 Apr	129—4	79—14			114—1			136—7			
7 May	130—8	69—16			102—7			140—4			
21 May	128—10	64—12					167—19	180—13			
4 Jun	129—14						143—1		149—3		
18 Jun	142—6					69—17	132—0		169—16		
2 Jul	112—11					128—5	132—16		186—16		
16 Jul	164—7						109—9		214—1	109—9	
31 Jul	152—10					139—15			181—0	113—10	153—10
13 Aug	143—14					78—9			124—16	47—10	139—0
27 Aug	129—15					66—16			142—16	78—9	106—4
10 Sep	130—3	8—0				130—1			200—5	170—9	123—4
24 Sep	126—9	29—17				122—8			177—10	135—19	194—16
8 Oct	131—0	64—19	41—16						178—17	156—17	146—16
22 Oct	119—0	87—12							185—16	155—17	103—8
4 Nov	118—6	78—5							131—14	134—13	118—14
19 Nov	125—10	80—15							125—17	123—12	77—6
3 Dec	100—10	68—15	58—12						77—5	101—17	124—6
17 Dec	20—0	106—15	68—12						102—8	101—11	122—18
31 Dec	92—3	68—1	31—14						97—19	95—8	84—16
Total	3174—8	1067—4	200—14	428—9	330—3	1339—0	1326—17	1136—19	2385—19	1525—1	1494—18

NOTE: Buckett, Old Rise and New Rise production teams paid at 1/4 per score
Goit ,, ,, ,, ,, ,, 0/10 per score

Appendix 4: CASH SALES BUXTON COLLIERIES 1790 (SCORE/CORVES)

Fortnight ending	Buckett Engine Pitt	Old Rise Pitt	New Rise Pitt	Goit
15 Jan	129—20	99—0		4—9
29 Jan	144—0	112—9		5—16
12 Feb	144—0			1—15
26 Feb	144—0			2—4
12 Mar	144—0			7—6
26 Mar	100—17			4—5
9 Apr	128—8	77—17		5—1
23 Apr	144—0	86—17		17—1
7 May	144—0	75—19		22—1
21 May	144—0	70—19		20—7
4 Jun	144—0			41—15
18 Jun	156—0			42—1
2 Jul	132—0			34—3
16 Jul	180—0			27—5
30 Jul	160—0			15—15
13 Aug	159—2			6—4
27 Aug	144—1			22—14
10 Sep	144—0	9—18		11—15
24 Sep	141—2	32—17		30—0
8 Oct	144—2	71—2	46—3	14—12
22 Oct	132—0	95—14		13—16
5 Nov	132—0	85—17		15—1
19 Nov	138—0	88—0		16—16
3 Dec	113—14	76—1	64—3	4—4
17 Dec	24—0	115—17	75—4	—
31 Dec	103—0	74—12	34—17	—
TOTAL	3513—16	1168—5	220—7	386—6

NOTE: Buckett, Old Rise and New Rise sold at 4/2 per score
Goit sold at 2/- per score

Appendix 5: CASH INCOME AND EXPENDITURE BUXTON COLLIERIES 1790 (£-s-d)

Fortnight ending	TOTAL CASH INCOME	CASH PAYMENTS Buckett Engine Pitt	Old Rise Pitt	New Rise Pitt	Goit	TOTAL CASH EXP'TURE
15 Jan	47- 3- 6	16-11- 3	7- 9- 3		19-10- 6	43-11- 1½
29 Jan	54- 0- 2	18-10- 6½	8- 3- 1		19-16-10	49-19- 5½
12 Feb	30- 3- 6	23- 7-11½	2- 6- 6		23- 9-11	49- 4- 4½
26 Feb	30- 4- 5½	22- 3-11			20-10- 9	41-17- 4
12 Mar	30-14- 7½	16- 6- 0		6- 5- 0	22-15- 3	45- 6-11½
26 Mar	23- 2- 0½	16- 3- 9			15-15- 9	31-19- 6
9 Apr	43- 9- 5½	13- 1- 9½	6- 5- 9	5- 2-10	15-17- 7	40- 7-11½
23 Apr	49-16- 0	17- 0-10	6-11- 1		14-12- 2	38-12- 1
7 May	52- 3- 9	19- 6- 3½	5-17-10		15-13- 7	56-14- 3
21 May	47-12- 6	16- 6-11	5- 8-11	11-10- 0	14-17- 6	75-10- 6
4 Jun	34- 3- 6	16-10- 7			21-12- 7½	52-14- 7½
18 Jun	36-14- 1	18- 9- 9			21- 1- 2½	53- 8- 1½
2 Jul	30-12- 4	16- 7- 6½			24-15- 4½	131- 5- 4
16 Jul	40- 4- 6	22-10- 7			33-17- 3	71-18- 1
30 Jul	36-11- 6	18-11- 7			37- 0- 8	55-12- 3
13 Aug	33-14-11	18-16- 4½			24-14- 3	43- 9- 7½
27 Aug	32- 5- 7	17- 1- 2½			35-18- 3½	62-14-11
10 Sep	33- 4- 9	19-13- 1½	1-12- 0		33-18- 9	55- 5- 6½
24 Sep	39- 4- 9½	15- 4- 7½	6- 1- 5½		33- 8- 0	55-19- 3½
8 Oct	55-10- 2½	16-10-10	5-11- 5	5-12- 0	27-16- 1	62- 1- 9
22 Oct	48-16- 4½	16- 5- 3	7-19- 7		24-14- 0½	58- 9-10
5 Nov	46-17- 9½	12-19- 3½	6- 9- 2		22- 8- 1½	43- 1- 3
19 Nov	48-15- 3½	24-16- 0	6-12- 6		19-11-10½	51- 0- 4½
3 Dec	53- 9- 9	11-19- 7	5-16- 0	5- 2- 5	19- 9-11½	42- 8- 0
17 Dec	44-16- 0½	12-17- 6½	8- 7- 2	5-16- 3½	20- 2- 0	47- 3- 0
31 Dec	44- 5- 2½	10-16- 1	5-12- 0½	3- 3- 1	16-12-11½	36- 4- 2
	1067-14- 7½	448- 9- 2½	96-14- 9	43- 1- 8	599-13- 0	1394-19- 7½

NOTE: The 'total' column for cash payments includes extra items not attributed to the individual pits

Appendix 6:

Conversion tables

a. Money

1d	=	½p
2d	=	1p
3d	=	1p
4d	=	1½p
5d	=	2p
6d	=	2½p
7d	=	3p
8d	=	3½p
9d	=	4p
10d	=	4p
11d	=	4½p
12s or 1s or 1/-	=	5p
2s or 2/-	=	10p
3s or 3/-	=	15p
4s or 4/-	=	20p
5s or 5/-	=	25p
6s or 6/-	=	30p
7s or 7/-	=	35p
8s or 8/-	=	40p
9s or 9/-	=	45p
10s or 10/-	=	50p
11s or 11/-	=	55p
12s or 12/-	=	60p
13s or 13/-	=	65p
14s or 14/-	=	70p
15s or 15/-	=	75p
16s or 16/-	=	80p
17s or 17/-	=	85p
18s or 18/-	=	90p
19s or 19/-	=	95p
20s or £1	=	100p or £1

b. Length

12 inches	=	1 foot
1 yard	=	0.9144 metre = 3 feet
1 metre	=	39.37 inches
22 yards	=	1 chain
10 chains	=	1 furlong
8 furlongs	=	1 mile
1 mile	=	1.61 kilometres = 1,760 yards
1 kilometre	=	0.62 mile

c. Area

9 square feet	4	1 square yard
30¼ square yards	4	1 perch
40 perches	4	1 rood
4 roods	4	1 acre
1 acre	4	0.405 hectare = 4,840 square yards
1 hectare	4	2.47 acres

d. Weight

1cwt.	=	50.8 kilograms
20cwt.	=	1 ton
1 ton		1016.05 kilograms

Appendix 7:

Miners named in Thomas Wyld's accounts 1790

NAME		TYPICAL TASK	DAY RATE (OR PIECE WORK)
Ashmore	Anthony	surface maintenance	1/6
,,	Anthony	driving the engine horse	6d (lad)
,,	Joshua	driving the engine horse	6d (lad)
,,	Thomas	coal getting	piece work
Bagshaw	Charles	coal getting	piece work
,,	Edward	delivering timber	1/6
,,	Robert	coal getting	piece work
Ball	John	coal preparation (scutching)	1/-
Barber	John	shaft sinking	2/-
Belford	James	coal getting	piece work
Bennett	John	banking	1/6
,,	Joshua	underground maintenace	1/2
,,	Samuel	banking	1/6
,,	William	banking	1/9
Brindley	John	getting stone	1/6
,,	William	getting stone	1/6
Brocklest	Joseph	shaft sinking	
		underground maintenance	1/2
Brown	John	coal getting	piece work
,,	Joshua	underground maintenance	1/-
Clewes	James	delivering lime	2/-
Dixon	John	underground maintenance	1/8
Goodwin	Jove	getting stone	1/6
,,	Thomas	winding coal	piece work
Greenhough	Edward	building a bridge	2/-
,,	Isaac	shaft sinking	2/-
Heald	Thomas	shaft sinking	piece work
		underground maintenance	1/6
Hibbert	John	coal preparation	1/4
Hudson	James	coal getting	piece work
Kirk	George	underground maintenance	2/-
,,	Nicholas	coal getting	piece work
,,	Thomas	filling coal	1/10
,,	William	boating the coal	8d (lad)
Lomas	Isaac	underground development	1/6
Longson	William	getting stone, repairing roads	1/2
Nadin	John	shaft sinking	piece work
,,	Thomas	banking	1/6
,,	William	underground development	1/2
Norton	John	driving the engine horse	6d (lad)
,,	William	boating the coal	1/- (lad)
Plant	Gilbert	guttering	piece work
,,	John	surface maintenance	1/6
Renshaw	John	carpenter	1/6
,,	Joshua	boating the coal	8d (lad)
Salt	John	building a bridge	2/-
Slater	George	surface maintenance	1/6
Street	John	underground development	1/6

Sutton	James	winding coal	piece work
,,	John	surface maintenance	1/6
,,	Thomas	shaft sinking	2/-
Tamblin	Joshua	coal getting	piece work
Taylor	John	surface maintenance	piece work
Trafford	Samuel	surface maintenance	piece work
,,	William	underground development	1/2
Turner	George	driving the engine horse	6d (lad)
,,	Edward	underground development	1/6
,,	Peter	driving the engine horse	6d (lad)
,,	Phillip	delivering timber	1/6
,,	Samuel	banking	1/6
Vawdrey	William	coal getting	piece work
Ward	Edward	coal preparation	1/- — 1/6
,,	John	banking	1/6
,,	Samuel	coal getting	piece work
Warren	William	underground development	1/6
Wheeldon	David	driving the engine horse	6d (lad)
,,	Jacob	boating the coal	8d (lad)
,,	John	underground maintenance	1/2
,,	Thomas	driving the engine horse	6d (lad)
,,	Mathias	driving the engine horse	6d (lad)
Wild	John	underground maintenance	1/2
,,	Thomas	'clerk of works'	salary
,,	William	coal getting	piece work
Yates	George	shaft sinking	1/8

Appendix 8: Coal production 1859-1902 as recorded on a plan held by the East Midlands Mining Records Office

YEAR to 30 June	AREA—Acres—Roods—Perches Goyt	House Coal	Estimated tonnages Goyt	House Coal	Total
(1/1/59-30/6/62)	4-2-7	3-2-5	22700	17700	40400
1863	2-3-16	1-1-5	14200	6400	20600
1864	4-0-10	0-2-11	20300	2800	23100
1865	4-2-23	1-0-12	23200	5400	28600
1866	4-0-30	0-3-27	21000	4600	25600
1867	4-3-25	1-1-13	24500	6600	31100
1868	4-3-2	1-0-2	23800	5100	28900
1869	5-1-17	0-1-17	26800	1800	28600
1870	4-2-27	0-3-27	23300	4600	27900
1871	5-3-12	0-1-31	29100	2200	31300
1872	5-0-26	0-2-15	25800	3000	28800
1873	3-3-25	1-1-37	19400	7400	26800
1874	3-3-18	1-1-18	19200	6800	26000
1875	4-2-33	0-2-30	23500	3400	26900
1876	5-1-5	0-3-7	26400	4000	30400
1877	5-0-20	0-1-8	25600	1500	27100
1878	5-0-3	0-1-26	25100	2000	27100
1879	4-1-36	0-1-20	22400	1900	24300
1880	4-0-36	0-2-9	21100	2800	23900
1881	3-3-27	0-3-3	19500	3800	23300
1882	4-0-1	0-1-26	20000	2000	22000
1883	3-0-29	0-2-16	15900	3000	18900
1884	4-2-1	0-0-39	22500	1200	23700
1885	3-1-3	0-1-37	16300	2400	18700
1886	3-2-39	0-1-9	18700	1500	20200
1887	4-0-24	0-2-23	20700	3200	23900
1888	4-0-11	0-2-19	20300	3100	23400
1889	1-3-18	0-2-4	9300	2600	11900
1890	1-3-5	1-0-7	8900	5200	14100
1891	1-0-16	0-3-31	5500	4700	10200
1892	2-0-5	0-3-34	10100	4800	14900
1893	4-2-8	0-2-11	22700	2800	25500
1894	0-0-39	0-2-30	1200	3400	4600
1895	0-0-39	0-2-8	1200	2700	3900
1896	0-0-39	0-2-7	1200	2700	3900
1897	0-0-39	0-1-2	1200	1300	2500
1898	0-0-39	0-0-39	1200	1200	2400
1899	0	0-0-26	0	800	800
1900	0	0-2-0	0	2500	2500
1901	0	0-0-37	0	1100	1100
1902	0	0-0-31	0	1000	1000
			653800	147000	800800
	Estimated 1903-1919		0	20000	20000
		TOTAL	652800	167000	820800

N.B. The plan records the area of coal worked in each year to 1902. These figures have been converted to tonnages by taking 1 Acre as equivalent to 5000 tons of coal production. Estimates from 1903-1919 have been made directly from the plan.

Appendix 9:

Information from Census Returns: 1840 (Place of birth not given)

	Name	Age	Description	Address
Ashmore	George	35	collier	Gate Farm
,,	James	22	collier	Dog Hole
,,	John	27	collier	Dog Hole
,,	Thomas	35	collier	Germany
,,	William	17	collier	Dog Hole
,,	—	58	collier	Dog Hole
Bagshaw	Charles	15	collier	Shay Lodge
,,	Charles	12	collier	,,
,,	James	15	collier	,,
,,	Joseph	40	collier	,,
,,	Samuel	20	collier	,,
,,	Thomas	16	collier	,,
Bennett	George	22	collier	Handcroft
,,	John	30	collier	Handcroft
,,	Peter	25	collier	(illegible)
,,	Richard	30	collier	,,
,,	Richard	50	collier	Handcroft
,,	William	31	collier	Handcroft
Hull	John	20	coal miner	Nether Green
,,	William	75	coal miner	Nether Green
Hutton	James	37	collier	Handcroft
Oliver	Thomas	55	collier	Brandside
Read	William	35	collier	Washgate
Ward	Edward	35	banksman	Goyt Moss
,,	John	55	collier	Goyt Moss
,,	Samuel	52	collier	Gutter House
Street	Edward	15	collier	Level House
,,	John	55	collier	Level House

Appendix 10:

Information from Census Returns: 1851

	Name	Age	Description	Address	Place of birth
Ashmore	George	35	collier	Burbage	Burbage
	James	32	collier	Axe Edge	Burbage
	James	70	coal carrier	Cote Heath	Buxton
	John	37	coal miner	Wye Head	Burbage
	Thomas	46	collier	Burbage	Burbage
	William	27	collier	,,	Burbage
	,,	18	miner	,,	Burbage
	,,	14	working in coal mine	Wye Head	Burbage
Bagshaw	Charles	27	collier	Handcroft	Burbage
Bennett	Edward	17	collier	Handcroft	Burbage
	Joseph	14	,,	,,	Burbage
	Samuel	19	,,	,,	Burbage
	William	41	collier	,,	Burbage
	William	22	miner	,,	Burbage
Gregory	William	54	banksman	Green Lane	Buxton
Holme	William	46	collier	Grin	Burbage
Hall	William	31	miner	Wye Head	Burbage
Norton	Josh	16	coal carrier	Burbage	Burbage
	Samuel	12	,,	,,	Burbage
	William	14	,,	,,	Burbage
Street	Edward	26	banksman	Level	Buxton
	Robert	36	collier	Burbage	Burbage
	John	66	banksman	Level	Buxton
Ward	Samuel	58	collier	Gutter	Burbage
	David	22	collier	,,	Burbage
	David	51	banksman	Goits Moss	Macclesfield Forest
	John	68	collier	Liandry House	Burbage
Wharmby	James	20	collier	Level Colliery	Buxton
	John	15	,,	,,	Buxton
	Samuel	13	,,	,,	Buxton
	William	48	banksman	,,	New Mills

Note As there were also barytes (caulk) and lead mines in the district at this time the 'miners' may be connected with these: one lead miner is recorded as such — a Mr. Doxey, born in Middleton (near Wirksworth).

Appendix 11: Information from Census Returns: 1881

	Name	Age	Description	Address	Place of birth
Ashmore	John	39	coal miner	Hill Top, Axe Edge	Burbage
,,	Thomas	41	coal miner	Gate Farm	Burbage
,,	Thomas	39	coal miner	Germany	Burbage
,,	Thomas	54	coal miner	Macclesfield Old Road	Burbage
Bagshaw	Charles	54	coal miner	Shay Lodge	Burbage
,,	Thomas	27	coal miner	Engine House	Burbage
Bennett	Joseph	44	coal miner	Level Cottages	Burbage
,,	Samuel	27	coal miner	Level Cottages	Hartington U.Q.
,,	Thomas	47	colliery labourer	Dog Hole	Burbage
,,	William	71	unemployed coal miner	Level Cottages	Hartington U.Q.
Brown	George	39	coal miner	Green Lane	Norbury, Staffs.
Bowyer	James	60	coal miner	Boothman's Cottages	Leek
,,		20	coal miner	,,	Heathcote, Staffs.
,,		18	coal miner	,,	,,
,,		16	coal miner	,,	,,
Brown	George	39	coal miner	Green Lane	Norbury, Staffs.
Brunt	George	16	Waggoner in coal mines	Canhole Cottages	Wildboarclough
Day	William	50	manager, coal mines	Macclesfield Old Road	Heanor
Goodwin	Abraham	60	coal miner	Grin Row	Flash
,,	John	15	coal miner's assistant	Macclesfield Old Road	Hartington U.Q.
Hadfield	Francis	46	coal miner	Level Cottages	Whaley Bridge
,,	Francis	20	coal miner	,,	Burbage
Hay	John	24	colliery engineer	Macclesfield Old Road	Heanor
Heape	George	35	collier	Hill Top, Axe Edge	Quarnford
Heywood	Samuel	30	coal miner	Level Cottages	Pott Shrigley
Mellor	John	46	coal miner	Lime Row	Staffs.
Morton	William	47	coal miner	Green Lane	Aston on Trent
Nadin	William	24	coal miner	Handcroft	Macclesfield
Needham	James	25	coal miner	Macclesfield New Road	Whaley Bridge
Norton	Francis	23	coal miner	Green Lane	Heanor
Packwood	George	37	coal miner	Harper Hill	Worcestershire
Pine	John	27	coal miner	Rose Cottages	Devonshire
Price	John	57	coal miner	Macclesfield Old Road	Okehampton
Robinson	John	17	coal miner	Rose Cottages	Flash
Salt	John	40	coal miner	Macclesfield Old Road	Quarnford
Shufflebotham	William	26	coal miner	Boothman's Cottages	Quarnford
Street	William	22	coal miner	Level Cottages	Burbage

FOOTNOTES

Chapter One

1. R. Grundey Heape. Buxton under the Dukes of Devonshire. 1948.

Chapter Two

1. J. Nef. The Rise of the British Coal Industry. Vol. I. 1932.
2. Ibid. Vol. I. p. 158.
3. Ibid. Vol. I. p. 19.
4. John Farey. General View of the Agriculture and Minerals of Derbyshire. Vol. I. (Originally published in 1811.)
5. J. Nef. The Rise of thc British Coal Industry. Vol. I. p. 61.
6. John Farey. Vol. I. pp. 188-215. The list of collieries was originally published in the Philosophical Magazine, Vol. 35. pp. 431-438. 1810.
7. Thomas Short. The Natural, Experimental and Medicinal History of the Mineral Waters of Derbyshire, Lincolnshire and Yorkshire. 1734.
8. White Watson. Strata of Derbyshire. 1811. pp. 22-23.

Chapter Three

1. Dodd and Dodd. Peakland Roads and Trackways. 1980. p. 97. Dodd and Dodd refer to Jaggers Gate being on the line of the 1759 Macclesfield-Buxton turnpike. Evidence on the ground shows the turnpike to be of much later date.
2. Derbyshire Record Office. Q/RIc32.
3. SK 132.615.

Chapter Four

1. Celia Fiennes. Through England on a Side-saddle.
2. Helen Harris. Industrial Archaeology of the Peak District. 1971. p. 61. L. Jackson. The Buxton Lime Trade. 1963. p. 9. Reprinted from Cement and Gravel. 1950.
3. Ibid. p. 12. Pilkington. A View of the Present State of Derbyshire 1789. Vol. II. p. 292.
4. Buxton Museum and Art Gallery.
5. Thomas Short. op. cit. p. 25.
6. Chatsworth. Ref. L/76/20.
7. SK 119.656.
8. Farey. op. cit. Vol. II. pp. 436-437.
9. Ibid. Vol. I. pp. 188-215.
10. Ibid. Vol. II. p. 435.
11. Ibid. Vol. II. p. 440.

Chapter Five

1. R. L. Galloway. Annals of Coal Mining and the Coal Trade. 1898. Reprint 1971. Vol. I. p. 61.
2. Nef. op. cit. p. 61.
3. Public Record Office. MPC 274(1).
4. John Ogilby. Britannia Depicta. 1675.
5. L. Jackson. The Buxton Lime Trade. 1963.
6. Thomas Short. op. cit.

Chapter Six

1. Chatsworth. Ref. L/110/47.
2. Nef. op. cit. Vol. II. p. 375.
3. Reference: Mr. A. Poulson, who worked in the mines in the Pott Shrigley area until their closure.
4. Mine abandonment plans 3072 and 6915. Mining Record Office, Eastwood.
5. OS 25in. map. Sheet XXI. 1880.
6. J. A. Robey and L. Porter. The Copper and Lead Mines of Ecton Hill, Staffordshire. 1972.
7. Galloway. op. cit. Vol. I. p. 271.
8. I. Hall. Georgian Buxton. 1984. p. 23.

Chapter Seven

1. Mine abandonment plan 6915. Mining Record Office, Eastwood.
2. Farey. op. cit. Vol. I.
3. Mine abandonment plan 3072. Mining Record Office, Eastwood.
4. Reference: Mr. J. Bagshaw.
5. Jackson. op. cit.
6. Chatsworth. Buxton Estate Office Accounts.
7. Mineral Statistics of the U.K.
8. J. Marshall. The Cromford and High Peak Railway. 1982. p. 17.
9. Jackson. op. cit.
10. Farey. op. cit. Vol. I. p. 216.
11. Stephen Glover. History and Gazetteer of Derbyshire. 1831. p. 45.
12. Reference: Mrs. A. Phillips, who carted coal from Cistern's Clough for 6½ years prior to the colliery's closure.
13. F. W. Cope. Correlation of the coal measures of Macclesfield and the Goyt trough. Transactions: Institution of Mining Engineers.
14. SK 1725.6815.

Chapter Eight

1. Farey. op. cit.
2. Chatsworth. Buxton Estate Office Accounts.
3. Reference: Mrs. A. Phillips. See Chap. Seven, Note 12.

Chapter Nine

1. Dr. Robertson. Guide to Buxton. Eleventh edition. 1886.
2. The Reliquary. Vol. 6. p. 150.
3. Notes of the Peakland Archaeological Society.
4. Burt, Atkinson, Waite and Burnley. Derbyshire Mineral Statistics 1845-1913. 1981. p. 20.

Chapter Ten

1. OS 6in. map. Sheet XXI.4. 1880.

www.ingramcontent.com/pod-product-compliance
Ingram Content Group UK Ltd.
Pitfield, Milton Keynes, MK11 3LW, UK
UKHW020139250726
13967UKWH00002B/748